LONDON MATHEMATICAL SOCIETY LECTURE NOTE SERIES

Managing Editor: Professor J.W.S. Cassels, Department of Pure Mathematics and Mathematical Statistics, University of Cambridge, 16 Mill Lane, Cambridge CB2 1SB, England

The titles below are available from booksellers, or, in case of difficulty, from Cambridge University Press.

46 p-adic analysis: a short course on recent work, N. KOBLITZ
50 Commutator calculus and groups of homotopy classes, H.J. BAUES
59 Applicable differential geometry, M. CRAMPIN & F.A.E. PIRANI
66 Several complex variables and complex manifolds II, M.J. FIELD
69 Representation theory, I.M. GELFAND *et al*
86 Topological topics, I.M. JAMES (ed)
87 Surveys in set theory, A.R.D. MATHIAS (ed)
88 FPF ring theory, C. FAITH & S. PAGE
89 An F-space sampler, N.J. KALTON, N.T. PECK & J.W. ROBERT
90 Polytopes and symmetry, S.A. ROBERTSON
92 Representation of rings over skew fields, A.H. SCHOFIELD
93 Aspects of topology, I.M. JAMES & E.H. KRONHEIMER (eds)
94 Representations of general linear groups, G.D. JAMES
96 Diophantine equations over function fields, R.C. MASON
97 Varieties of constructive mathematics, D.S. BRIDGES & F. RICHMAN
98 Localization in Noetherian rings, A.V. JATEGAONKAR
99 Methods of differential geometry in algebraic topology, M. KAROUBI & C. LERUSTE
100 Stopping time techniques for analysts and probabilists, L. EGGHE
104 Elliptic structures on 3-manifolds, C.B. THOMAS
105 A local spectral theory for closed operators, I. ERDELYI & WANG SHENGWANG
107 Compactification of Siegel moduli schemes, C.-L. CHAI
109 Diophantine analysis, J. LOXTON & A. VAN DER POORTEN (eds)
110 An introduction to surreal numbers, H. GONSHOR
113 Lectures on the asymptotic theory of ideals, D. REES
114 Lectures on Bochner-Riesz means, K.M. DAVIS & Y.-C. CHANG
116 Representations of algebras, P.J. WEBB (ed)
118 Skew linear groups, M. SHIRVANI & B. WEHRFRITZ
119 Triangulated categories in the representation theory of finite-dimensional algebras, D. HAPPEL
121 Proceedings of *Groups - St Andrews 1985*, E. ROBERTSON & C. CAMPBELL (eds)
122 Non-classical continuum mechanics, R.J. KNOPS & A.A. LACEY (eds)
128 Descriptive set theory and the structure of sets of uniqueness, A.S. KECHRIS & A. LOUVEAU
129 The subgroup structure of the finite classical groups, P.B. KLEIDMAN & M.W. LIEBECK
130 Model theory and modules, M. PREST
131 Algebraic, extremal & metric combinatorics, M.-M. DEZA, P. FRANKL & I.G. ROSENBERG (eds)
132 Whitehead groups of finite groups, ROBERT OLIVER
133 Linear algebraic monoids, MOHAN S. PUTCHA
134 Number theory and dynamical systems, M. DODSON & J. VICKERS (eds)
135 Operator algebras and applications, 1, D. EVANS & M. TAKESAKI (eds)
137 Analysis at Urbana, I, E. BERKSON, T. PECK, & J. UHL (eds)
138 Analysis at Urbana, II, E. BERKSON, T. PECK, & J. UHL (eds)
139 Advances in homotopy theory, S. SALAMON, B. STEER & W. SUTHERLAND (eds)
140 Geometric aspects of Banach spaces, E.M. PEINADOR & A. RODES (eds)
141 Surveys in combinatorics 1989, J. SIEMONS (ed)
144 Introduction to uniform spaces, I.M. JAMES
145 Homological questions in local algebra, JAN R. STROOKER
146 Cohen-Macaulay modules over Cohen-Macaulay rings, Y. YOSHINO
148 Helices and vector bundles, A.N. RUDAKOV *et al*
149 Solitons, nonlinear evolution equations and inverse scattering, M. ABLOWITZ & P. CLARKSON
150 Geometry of low-dimensional manifolds 1, S. DONALDSON & C.B. THOMAS (eds)
151 Geometry of low-dimensional manifolds 2, S. DONALDSON & C.B. THOMAS (eds)
152 Oligomorphic permutation groups, P. CAMERON
153 L-functions and arithmetic, J. COATES & M.J. TAYLOR (eds)
155 Classification theories of polarized varieties, TAKAO FUJITA
156 Twistors in mathematics and physics, T.N. BAILEY & R.J. BASTON (eds)
158 Geometry of Banach spaces, P.F.X. MÜLLER & W. SCHACHERMAYER (eds)
159 Groups St Andrews 1989 volume 1, C.M. CAMPBELL & E.F. ROBERTSON (eds)
160 Groups St Andrews 1989 volume 2, C.M. CAMPBELL & E.F. ROBERTSON (eds)
161 Lectures on block theory, BURKHARD KÜLSHAMMER
162 Harmonic analysis and representation theory, A. FIGA-TALAMANCA & C. NEBBIA
163 Topics in varieties of group representations, S.M. VOVSI
164 Quasi-symmetric designs, M.S. SHRIKANDE & S.S. SANE
166 Surveys in combinatorics, 1991, A.D. KEEDWELL (ed)
168 Representations of algebras, H. TACHIKAWA & S. BRENNER (eds)

D1105007

FLORIDA STATE UNIVERSITY LIBRARIES

JUN 18 1997

TALLAHASSEE, FLORIDA

169 Boolean function complexity, M.S. PATERSON (ed)
170 Manifolds with singularities and the Adams-Novikov spectral sequence, B. BOTVINNIK
171 Squares, A.R. RAJWADE
172 Algebraic varieties, GEORGE R. KEMPF
173 Discrete groups and geometry, W.J. HARVEY & C. MACLACHLAN (eds)
174 Lectures on mechanics, J.E. MARSDEN
175 Adams memorial symposium on algebraic topology 1, N. RAY & G. WALKER (eds)
176 Adams memorial symposium on algebraic topology 2, N. RAY & G. WALKER (eds)
177 Applications of categories in computer science, M. FOURMAN, P. JOHNSTONE & A. PITTS (eds)
178 Lower K- and L-theory, A. RANICKI
179 Complex projective geometry, G. ELLINGSRUD *et al*
180 Lectures on ergodic theory and Pesin theory on compact manifolds, M. POLLICOTT
181 Geometric group theory I, G.A. NIBLO & M.A. ROLLER (eds)
182 Geometric group theory II, G.A. NIBLO & M.A. ROLLER (eds)
183 Shintani zeta functions, A. YUKIE
184 Arithmetical functions, W. SCHWARZ & J. SPILKER
185 Representations of solvable groups, O. MANZ & T.R. WOLF
186 Complexity: knots, colourings and counting, D.J.A. WELSH
187 Surveys in combinatorics, 1993, K. WALKER (ed)
188 Local analysis for the odd order theorem, H. BENDER & G. GLAUBERMAN
189 Locally presentable and accessible categories, J. ADAMEK & J. ROSICKY
190 Polynomial invariants of finite groups, D.J. BENSON
191 Finite geometry and combinatorics, F. DE CLERCK *et al*
192 Symplectic geometry, D. SALAMON (ed)
193 Computer algebra and differential equations, E. TOURNIER (ed)
194 Independent random variables and rearrangement invariant spaces, M. BRAVERMAN
195 Arithmetic of blowup algebras, WOLMER VASCONCELOS
196 Microlocal analysis for differential operators, A. GRIGIS & J. SJÖSTRAND
197 Two-dimensional homotopy and combinatorial group theory, C. HOG-ANGELONI, W. METZLER & A.J. SIERADSKI (eds)
198 The algebraic characterization of geometric 4-manifolds, J.A. HILLMAN
199 Invariant potential theory in the unit ball of $\mathbf{C}^n$, MANFRED STOLL
200 The Grothendieck theory of dessins d'enfant, L. SCHNEPS (ed)
201 Singularities, JEAN-PAUL BRASSELET (ed)
202 The technique of pseudodifferential operators, H.O. CORDES
203 Hochschild cohomology of von Neumann algebras, A. SINCLAIR & R. SMITH
204 Combinatorial and geometric group theory, A.J. DUNCAN, N.D. GILBERT & J. HOWIE (eds)
205 Ergodic theory and its connections with harmonic analysis, K. PETERSEN & I. SALAMA (eds)
206 An introduction to noncommutative differential geometry and its physical applications, J. MADORE
207 Groups of Lie type and their geometries, W.M. KANTOR & L. DI MARTINO (eds)
208 Vector bundles in algebraic geometry, N.J. HITCHIN, P. NEWSTEAD & W.M. OXBURY (eds)
209 Arithmetic of diagonal hypersurfaces over finite fields, F.Q. GOUVÊA & N. YUI
210 Hilbert C*-modules, E.C. LANCE
211 Groups 93 Galway / St Andrews I, C.M. CAMPBELL *et al*
212 Groups 93 Galway / St Andrews II, C.M. CAMPBELL *et al*
214 Generalised Euler-Jacobi inversion formula and asymptotics beyond all orders, V. KOWALENKO, N.E. FRANKEL, M.L. GLASSER & T. TAUCHER
215 Number theory 1992–93, S. DAVID (ed)
216 Stochastic partial differential equations, A. ETHERIDGE (ed)
217 Quadratic forms with applications to algebraic geometry and topology, A. PFISTER
218 Surveys in combinatorics, 1995, PETER ROWLINSON (ed)
220 Algebraic set theory, A. JOYAL & I. MOERDIJK
221 Harmonic approximation, S.J. GARDINER
222 Advances in linear logic, J.-Y. GIRARD, Y. LAFONT & L. REGNIER (eds)
223 Analytic semigroups and semilinear initial boundary value problems, KAZUAKI TAIRA
224 Computability, enumerability, unsolvability, S.B. COOPER, T.A. SLAMAN & S.S. WAINER (eds)
225 A mathematical introdcution to string theory, S. ALBEVERIO, J. JOST, S. PAYCHA, S. SCARLATTI
226 Novikov conjectures, index theorems and rigidity I, S. FERRY, A. RANICKI & J. ROSENBERG (eds)
227 Novikov conjectures, index theorems and rigidity II, S. FERRY, A. RANICKI & J. ROSENBERG (eds)
228 Ergodic theory of $\mathbf{Z}^d$ actions, M. POLLICOTT & K. SCHMIDT (eds)
229 Ergodicity for infinite dimensional systems, G. DA PRATO & J. ZABCZYK
230 Prolegomena to a middlebrow arithmetic of curves of genus 2, J.W.S. CASSELS & E.V. FLYNN
231 Semigroup theory and its applications, K.H. HOFMANN & M.W. MISLOVE (eds)
232 The descriptive set theory of Polish group actions, H. BECKER & A.S. KECHRIS
233 Finite fields and applications, S. COHEN & H. NIEDERREITER (eds)
234 Introduction to subfactors, V. JONES & V.S. SUNDER
235 Number theory 1993–94, S. DAVID (ed)
236 The James forest, H. FETTER & B. GAMBOA DE BUEN
237 Sieve methods, exponential sums, and their applications in number theory, G.R.H. GREAVES, G. HARMAN & M.N. HUXLEY (eds)
238 Representation theory and algebraic geometry, A. MARTSINKOVSKY & G. TODOROV (eds)
239 Clifford algebras and spinors, P. LOUNESTO
240 Stable groups, FRANK O. WAGNER
242 Geometric Galois Actions I, L. SCHNEPS & P. LOCHAK (eds)

London Mathematical Society Lecture Note Series. 234

Introduction to Subfactors

V. Jones
University of California, Berkeley

V.S. Sunder
Institute of Mathematical Sciences, Madras

Sci
QA
326
J66
1997

PUBLISHED BY THE PRESS SYNDICATE OF THE UNIVERSITY OF CAMBRIDGE
The Pitt Building, Trumpington Street, Cambridge CB2 1RP, United Kingdom

CAMBRIDGE UNIVERSITY PRESS
The Edinburgh Building, Cambridge, CB2 2RU, United Kingdom
40 West 20th Street, New York, NY 10011-4211, USA
10 Stamford Road, Oakleigh, Melbourne 3166, Australia

© V. Jones and V.S. Sunder 1997

This book is in copyright. Subject to statutory exception
and to the provisions of relevant collective licensing agreements,
no reproduction of any part may take place without
the written permission of Cambridge University Press.

First published 1997

Printed in the United Kingdom at the University Press, Cambridge

A catalogue record for this book is available from the British Library

ISBN 0 521 58420 5 paperback

Dedicated
to
Our wives

Contents

Preface

It must be stated at the outset that this little monograph has no pretensions to being a general all-purpose text in operator algebras. On the contrary, it is an attempt to introduce the potentially interested reader – be it a graduate student or a working mathematician who is not necessarily an operator algebraist – to a selection of topics in the theory of subfactors, this selection being influenced by the authors' tastes and personal viewpoints. For instance, we restrict ourselves to the theory of (usually hyperfinite) II_1 factors and their subfactors (almost always of finite index); thus, factors of type III do not make an appearance beyond the first (introductory) chapter, and the Tomita–Takesaki theorem makes only a cameo appearance in the appendix. It is hoped that such 'simplifications' will help to make the material more accessible to the uninitiated reader.

The aim of this book is to give an introduction to some of the beautiful ideas and results which have been developed, since the inception of the theory of subfactors, by such mathematicians as Adrian Ocneanu and Sorin Popa; an attempt has been made to keep the material as self-contained as possible; in fact, we feel it should be possible to use this monograph as the basis of a two-semester course to second year graduate students with a minimal background in Hilbert space theory.

A remark is in order, as far as the references are concerned; when we state certain standard facts without proof, the reader is often referred to a text in operator algebras; if, in this process, it seems that the text by the second author is cited more often than other texts, that is simply because, a reference for that exact fact already being known to exist at that particular place, it was possible to avoid a search for that fact in other texts.

We now give a brief outline of the contents of this volume for the sake of the possibly interested specialist.

The first chapter begins with a quick introduction to some preliminary facts about von Neumann algebras (Murray–von Neumann classification of factors and introduction to traces); this first section contains no proofs, but most of the sequel – with the notable exception of Popa's theorem on amenable subfactors – is self-contained modulo the unproved facts here. The next section starts with the *GNS* construction and goes on to discuss the *standard form* of a finite von Neumann algebra and, in particular, identifies the commutant of the left-regular representation with the range of the right-regular representation. The chapter continues with a discussion of crossed products by countable groups and concludes with some examples – the left von Neumann algebra of an *ICC* group, factors of the three types coming from crossed products of commutative von Neumann algebras with countable groups acting ergodically and freely, a model of the hyperfinite factor which demonstrates that its fundamental group is the entire positive line, and finally infinite tensor products and the definition of the hyperfinite factor.

The second chapter starts with the classification of (separable) modules over a von Neumann algebra (with separable pre-dual), continues with the definition and some elementary properties of the M-dimension of a (separable) module over a II_1 factor M, and concludes with the definition and some elementary properties of the index of a subfactor of a II_1 factor, the statement of the result on restrictions on index values and a proof of the fact that all index values in the interval $[4, \infty)$ are possible.

The third chapter begins with a section on the fundamental notion of the basic construction; the next section gathers together the basic facts about inclusions of finite-dimensional C^*-algebras (including the fact about the basic construction and 'reflection of Bratteli diagrams' as well as the notion of a Markov trace); the final section introduces the all-important sequence $\{e_n\}$, derives the basic properties of this sequence, and indicates the relation between the theorem on restrictions of index values and the classification of non-negative integral matrices of norm less than 2, as well as an outline of the procedure originally adopted to prove the existence of hyperfinite subfactors of index $4\cos^2\frac{\pi}{n}$.

The fourth chapter is devoted to the principal (or standard) graph invariant of a subfactor. It starts with a discussion of ('bifinite') bimodules over a pair of II_1 factors (contragredients and tensor products); the second section gives the two descriptions (in terms of the sequence of higher relative commutants as well as in terms of the bimodules that occur in the tower of the basic construction) of the principal graph of a subfactor, reduces the result on restriction of index values to the classification of non-negative integral matrices of small norm, and concludes with some examples of principal graphs; the chapter concludes with a discussion of 'Pimsner–Popa bases' and a proof of why the higher relative commutants have an interpretation as intertwiners of bimodules.

The fifth chapter starts with Pimsner and Popa's minimax characterisation of the index of a subfactor, the consequent estimation of the index of a hyperfinite subfactor in terms of an approximating 'ladder' of finite-dimensional C^*-algebras, and introduces the important notion of a commuting square; the second section discusses examples of commuting squares (vertex and spin models, and the braid-group example); the next section is devoted to the relation between commuting squares and the basic construction, and the consequent importance of symmetric or non-degenerate commuting squares (with respect to the Markov trace); the next two sections are devoted to the path-algebra model for a tower of finite-dimensional C^*-algebras, and the reformulation of the commuting square condition in terms of 'biunitarity', respectively; the sixth section is a discussion (without proofs) of the canonical commuting square associated to a subfactor and Popa's theorem on the completeness of this invariant; the final section of this chapter centres around

Ocneanu's compactness theorem and the prescription it provides for computing the higher relative commutants of a subfactor built from an arbitrary initial commuting square.

The sixth and final chapter is devoted to the rich class of examples of subfactors provided by the so-called vertex and spin models. This chapter systematically develops a diagrammatic formulation to discuss the higher relative commutants in these examples, and also shows how to push this diagrammatic formulation through for general commuting squares.

The book concludes with an appendix, which contains some facts used in the text – such as the non-existence of two-sided (algebraic) ideals in finite factors – as well as a computation of the principal and dual graphs of the 'subgroup-subfactor' and the original derivation of the one-variable polynomial invariant of knots.

Finally, the book comes equipped with such customary trappings as a bibliography, some remarks of a bibliographic nature, and one index containing both terms and symbols used.

Acknowledgement

These notes have a two-part genesis. To start with, the first author gave a series of about seven lectures at the centres of the Indian Statistical Institute at New Delhi and Bangalore during the winter of '92–'93; these lectures were 'panoramic' in nature and did not contain too many proofs (because of the obvious time constraints). Later, the second author visited the Research Institute of Mathematical Sciences at Kyoto for nine months during '94–'95, during which period he gave a course of about 25 lectures – which was an expanded version of the earlier abridged version, and which contained proofs of most assertions and was addressed to the graduate students in the audience. These notes were the by-product of those two courses of lectures.

The first (resp., second) author would like to express his gratitude to Professors Kalyan Sinha and V. Radhakrishnan, the Indian Statistical Institute, the Raman Research Institute and the National Board for Higher Mathematics in India (resp., Professor Huzihiro Araki and the Research Institute of Mathematical Sciences, Kyoto University) for the hospitality and the roles they played in making these visits possible.

Chapter 1

Factors

1.1 von Neumann algebras and factors

The fundamental notion is that of a von Neumann algebra. This is, typically, what may be called the 'symmetries of a group'. The precise way of saying this is that a (concrete) *von Neumann algebra* is nothing but a set of the form $M = \pi(G)'$ – where π is a unitary representation of a group G on a Hilbert space $\mathcal{H}$, and S' denotes, for S a subset of $\mathcal{L}(\mathcal{H})$ (the algebra of all bounded operators on $\mathcal{H}$), the commutant of S defined by $S' = \{x' \in \mathcal{L}(\mathcal{H}): x'x = xx' \text{ for all } x \text{ in } S\}$. In other words M is the set of intertwiners of the representation π: thus, $x \in M \Leftrightarrow x\pi(g) = \pi(g)x$ for all g in G.

The usual definition of a von Neumann algebra is: 'a self-adjoint subalgebra of $\mathcal{L}(\mathcal{H})$ satisfying $M = M''$' (where we write M'' for $(M')'$). This is equivalent to the definition we have chosen to give. Reason: if $M = \pi(G)'$, then M is a self-adjoint subalgebra and $M = M''$, since $S' = S'''$ for all $S \subseteq \mathcal{L}(\mathcal{H})$; conversely, if $M = M''$ is a self-adjoint subalgebra, we may set G equal to the unitary group of M' and appeal to the almost obvious fact (cf. [Sun1], Lemma 0.4.7) that G linearly spans M' (so that $G' = (M')'$).

The canonical commutative examples of abstract von Neumann algebras turn out to be $L^\infty(X, \mu)$, while the basic non-commutative example is $\mathcal{L}(\mathcal{H})$.

The fundamental 'double commutant theorem' of von Neumann states that a self-adjoint unital subalgebra M of $\mathcal{L}(\mathcal{H})$ is weakly closed – i.e. $\langle x_n\xi, \eta\rangle \to \langle x\xi, \eta\rangle \ \forall \xi, \eta \in \mathcal{H}, x_n \in M \ \forall n \Rightarrow x \in M$) (if and only if M is σ-weakly closed – see §A.1) if and only if $M = M''$.

The importance of the notion of a von Neumann algebra was recognised in 1936 by Murray and von Neumann (although of course, they called them 'rings of operators') – see [MvN1] – who also quickly realised that the 'building blocks' in the theory of von Neumann algebras were (what they, and people after them, called) *factors*. If $M = \pi(G)'$, then M is a factor precisely when the representation is 'isotypical'. (If G is compact and π is a strongly continous representation, this says that π is a multiple of an irreducible representation;

for general G, this says that any two non-zero subrepresentations of π admit non-zero subrepresentations which are equivalent.)

More precisely, a von Neumann algebra M is called a factor if it has trivial centre – i.e., $Z(M) = M \cap M' = \mathbb{C}\cdot 1$; and von Neumann proved – see [vN3] – that any von Neumann algebra is 'a direct integral of factors'.

Projections play a major role in the theory. It is true that any von Neumann algebra is the norm-closed linear span of its projections – i.e., elements p satisfying $p = p^* = p^2$. (Reason: if $M = L^\infty(X, \mu)$, this is because simple functions – i.e., finite linear combinations of characteristic functions of sets – are dense in $L^\infty(X, \mu)$. The preceding statement and the spectral theorem show that any bounded normal operator is norm-approximable by finite linear combinations of its spectral projections. This proves the assertion about general von Neumann algebras.)

Two projections e and f in a von Neumann algebra M are said to be Murray–von Neumann equivalent – written $e \sim f$ (or $e \sim f$ (rel M)) – if there exists (a partial isometry) u in M such that $u^*u = e$ and $uu^* = f$. It is not too hard to show that if (and only if) M is a factor, any two projections in M are comparable in the sense that one is Murray–von Neumann equivalent to a sub-projection of the other.

Factors were initially classified into three broad types by Murray and von Neumann, on the basis of the structure of the lattice $\mathcal{P}(M)$ of projections in M. The key notion they use is that of a finite projection; say that a projection $e \in \mathcal{P}(M)$ is finite if e is not equivalent to any proper sub-projection of e. It should be obvious that a minimal projection of M, should one exist, is necessarily finite. It should also be equally clear that any factor is one and exactly one of the types I–III as defined below.

DEFINITION 1.1.1 *(a) A factor M is said to be of type:*

(i) I, if there exists a non-zero minimal projection in M;

(ii) II, if M contains non-zero finite projections and if M is not of type I; and

(iii) III, if no non-zero projection in M is finite.

(b) A factor M is said to be finite if 1 *(the multiplicative identity of M, which always exists – cf. the definition of a concrete von Neumann algebra) is a finite projection in M; equivalently M is finite if M does not contain any non-unitary isometry.*

One of the basics facts about finite factors is contained in the following result; for a proof, see, for instance [Tak1], Theorem V.2.6.

PROPOSITION 1.1.2 *If M is a finite factor, then there always exists a unique faithful normal tracial state (henceforth abbreviated simply to 'trace') on M;*

i.e., there exists a unique linear functional τ on M such that:

(i) (trace) $\tau(xy) = \tau(yx)$;

(ii) (state) $\tau(x^*x) \geq 0$ and $\tau(1) = 1$;

(iii) (faithful) $\tau(x^*x) \neq 0$ if $x \neq 0$; and

(iv) (normal) τ is (σ-)weakly continuous.

Further, if $e, f \in \mathcal{P}(M)$, then

$$e \sim f \Leftrightarrow \tau(e) = \tau(f). \tag{1.1.1}$$

We conclude this section with the rudiments of this Murray–von Neumann classification of factors. Suppose then that M is a factor with separable predual. Then consider the above-mentioned three possibilities for the type of M:

(I) If M is a factor of type I, it turns out (and it is not hard to prove – cf. [Sun1], Exercise 4.3.1, for instance) that $M \cong \mathcal{L}(\mathcal{H})$ for some separable Hilbert space $\mathcal{H}$ and that M is finite precisely when $\mathcal{H}$ is finite-dimensional. In particular, any finite factor of type I is necessarily finite-dimensional. We say that the type I factor M is of type I_n if $\mathcal{H}$ is an n-dimensional Hilbert space, for $n = 1, 2, \cdots, \infty$. If M is of type $I_n, 1 \leq n < \infty$, then $M \simeq M_n(\mathbb{C})$ and the unique trace tr of Proposition 1.1.2 is just the usual matrix-trace after suitable normalisation: thus, $\mathrm{tr}((x_{ij})) = \frac{1}{n}\sum_{i=1}^{n} x_{ii}$.

(II) If M is a factor of type II, there are two possibilities, according to whether or not M is a finite factor in the sense of Definition 1.1.1(b). We say that a type II factor M is of type II_1 or of type II_∞ according to whether or not M is a finite factor.

Recall, from Proposition 1.1.2 that every II_1 factor is equipped with a faithful normal tracial state τ. It is true – for instance, see [Sun1], Proposition 1.3.14 – that if M is a II_1 factor, then $\{\tau(p) : p \in \mathcal{P}(M)\} = [0, 1]$. Thus the trace τ induces a bijection between the collection of Murray–von Neumann equivalence classes of projections in a II_1 factor and the continuum [0,1]. What attracted von Neumann to II_1 factors was the possibility of 'continuously varying dimensions'.

On the other hand, suppose M is a II_∞ factor. Fix an arbitrary finite projection $p_1 \in M$. It is true, then, that there exists – see, for instance, [Sun1] – a sequence $\{p_n : 1 \leq n < \infty\}$ of mutually orthogonal projections in M such that (i) $p_n \sim p_1(\mathrm{rel}\ M)\forall n$, and (ii) $\sum_n p_n = 1$. Pick a partial isometry $u_n \in M$ such that $u_n^* u_n = p_n, u_n u_n^* = p_1$. (We briefly digress to remark that it is true – see [Tak1], for instance – that if M is a von Neumann algebra, and if $p \in \mathcal{P}(M)$, then the so-called 'corner' of M defined by $M_p = pMp = \{pxp : x \in M\}$ is again a von Neumann algebra.) It then follows easily that M_{p_1} is a II_1 factor, and that the mapping $x \mapsto ((u_i^* x u_j))$ establishes an isomorphism of M onto $M_{p_1} \otimes \mathcal{L}(\ell^2)$. Now consider the map $\mathrm{Tr} : M_+ (= \{x \in M : x \geq 0\}) \to [0, \infty]$ defined by $\mathrm{Tr}\, x = \sum_n \mathrm{tr}_M(u_n x u_n^*)$. It is true of this map – see [vN1] – that

(i) $\{\mathrm{Tr}\, p : p \in \mathcal{P}(M)\} = [0, \infty]$;

(ii) if $p, q \in \mathcal{P}(M)$, then $p \sim q(\text{rel } M) \Leftrightarrow \text{Tr}\, p = \text{Tr}\, q$; and

(iii) if $p, q \in \mathcal{P}(M)$, and $pq = 0$, then $\text{Tr}\,(p+q) = \text{Tr}\, p + \text{Tr}\, q$.

Notice that this choice of Tr is so 'normalised' that it takes the value 1 at p_1; it is true that but for this possible choice in scaling, the function Tr is uniquely determined by the above properties. This function Tr is referred to as the faithful normal semifinite trace on the II_∞ factor.

(III) The factor M is of type III precisely when every non-zero projection is infinite; under our standing assumption on the separability of the pre-dual of M, it turns out – cf. [Sun1], Corollary 1.2.4(b) – that any two infinite projections are equivalent. Thus M is a type III factor precisely when any non-zero projection is Murray–von Neumann equivalent to the identity 1.

1.2 The standard form

Suppose φ is a *normal state* on a von Neumann algebra M – i.e., φ is a linear functional on M which is:

(i) positive – i.e., $\varphi(x^*x) \geq 0$;

(ii) a state – $\varphi(1) = 1$;

(iii) normal – i.e., φ is σ-weakly continuous.

Consider the sesquilinear form on M defined by $(x, y) \to \langle x, y \rangle = \varphi(y^*x)$. This satisfies all the requirements of an inner product except positive definiteness – i.e., the set $N_\varphi = \{x \in M : \varphi(x^*x) = 0\}$ may be non-trivial. It is, in any case, a consequence of the Cauchy–Schwarz inequality that N_φ is a left-ideal of M. Hence the form $\langle \cdot, \cdot \rangle$ descends to a genuine inner product on the quotient space M/N_φ; further, the equation $\pi(x)(y + N_\varphi) = xy + N_\varphi$ defines, not just a well-defined, but even a bounded – with respect to the norm $||y + N_\varphi||_2 = \varphi(y^*y)^{1/2}$ – linear operator $\pi(x)$ on M/N_φ, which hence extends to the completion $\mathcal{H}_\varphi$. It is painless to verify that if $\pi_\varphi(x)$ denotes the extension to $\mathcal{H}_\varphi$ of $\pi(x)$, then π_φ defines a normal representation of M on $\mathcal{H}_\varphi$. Further, if $\xi_\varphi = 1 + N_\varphi$, then ξ_φ is a cyclic vector for π_φ, i.e., $[\pi_\varphi(M)\xi_\varphi] = \mathcal{H}_\varphi$ (where $[S]$ denotes the closed subspace spanned by a subset S of Hilbert space), and the given state φ is recovered from the triple $(\mathcal{H}_\varphi, \pi_\varphi, \xi_\varphi)$ by

$$\varphi(x) = \langle \pi_\varphi(x)\xi_\varphi, \xi_\varphi \rangle, x \in M.$$

We summarise the foregoing construction – called the *GNS* construction after Gelfand, Naimark and Segal – in the following:

PROPOSITION 1.2.1 *Let φ be a normal state on a von Neumann algebra M. Then there exists a triple $(\mathcal{H}, \pi, \xi)$ consisting of a Hilbert space $\mathcal{H}$ carrying a normal representation π of M, and a distinguished cyclic vector ξ of the representation satisfying $\varphi(x) = \langle \pi(x)\xi, \xi \rangle$ for all x in M.* □

Notice, incidentally, that if the state φ is faithful (meaning $\varphi(x^*x) \neq 0$ if $x \neq 0$), so is the representation π.

EXAMPLE 1.2.2 *If $M = L^\infty(X, \mu)$, a typical normal state on M is of the form $\phi_\nu(f) = \int f d\nu, f \in L^\infty(X, \mu)$ – where ν is a probability measure on X which is absolutely continuous with respect to μ. One GNS triple associated to ϕ_ν is given by $\mathcal{H}_\nu = L^2(X, \nu), \pi_\nu(f)\xi = f\xi$, and ξ_ν is the constant function* 1.
$\square$

It must be clear that a necessary and sufficient condition for a normal representation π of M to occur as a GNS representation (i.e., to be unitarily equivalent to one) is that the representation π be a cyclic representation.

An easy application of Zorn's lemma shows now that every (separable) normal representation of a von Neumann algebra is a (countable) direct sum of GNS representations.

In the rest of this section, we consider the special case when M admits a faithful normal tracial state, and analyse the GNS construction.

Assume thus that there exists a faithful normal tracial state τ on M. We write $L^2(M, \tau)$ for the Hilbert space underlying the GNS representation associated with τ. This representation is faithful, since $\pi(x) = 0 \Rightarrow 0 = \|\pi(x)\xi_\tau\|^2 = \tau(x^*x) \Rightarrow x = 0$. Hence we may – and do – identify x with $\pi_\tau(x)$ and we assume that $M \subseteq \mathcal{L}(L^2(M, \tau))$ so that there exists a cyclic vector Ω (which is the notation we shall employ for what we earlier called ξ_τ) such that $\tau(x) = \langle x\Omega, \Omega\rangle$ for all $x \in M$.

Besides being a cyclic vector for M, the vector Ω is also separating for M in the sense that $x\Omega = 0 (\Rightarrow \tau(x^*x) = \|x\Omega\|^2 = 0) \Rightarrow x = 0$.

Hence the Hilbert space $L^2(M, \tau)$ contains a vector Ω which is simultaneously cyclic and separating for M. We need the fact – which is ensured by the next lemma – that Ω is also a cyclic and separating vector for M'.

LEMMA 1.2.3 *If $M \subseteq \mathcal{L}(\mathcal{H})$ is a von Neumann algebra, a vector ξ in $\mathcal{H}$ is cyclic for M if and only if ξ is separating for M'.*

Proof : Let $\xi \in \mathcal{H}$. Let p' be the projection onto $[M\xi]$, the closed M-cyclic subspace spanned by ξ. Note that $p' \in M'$ and that $(1 - p')\xi = 0$. So, if ξ is separating for M', then $p' = 1$ so ξ is cyclic for M.

Conversely if ξ is cyclic for M, then $x' \in M'$ and $x'\xi = 0 \Rightarrow x'\eta = 0$ for all η in $[M\xi]$, since $x'(x\xi) = x(x'\xi) = 0$ for x in M. $\square$

Thus $L^2(M, \tau)$ has the vector Ω which is cyclic and separating for M as well as for M'. For any vector ξ in L^2, consider the two operators defined by:

$$\pi_\ell(\xi)(x'\Omega) = x'\xi \quad \forall x' \in M', \tag{1.2.1}$$
$$\pi_r(\xi)(x\Omega) = x\xi \quad \forall x \in M. \tag{1.2.2}$$

Thus, dom $\pi_\ell(\xi) = M'\Omega$ (and dom $\pi_r(\xi) = M\Omega$); these operators are densely and unambiguously defined since Ω is a cyclic and separating vector for M and for M'.

Call a vector ξ *left-bounded* (respectively *right-bounded*) if the operator $\pi_\ell(\xi)$ (resp., $\pi_r(\xi)$) extends to a (necessarily unique) bounded operator on all of $\mathcal{H}$ (or equivalently, is bounded on its dense domain of definition).

If ξ is left- (resp. right-) bounded, we shall continue to write $\pi_\ell(\xi)$ (resp. $\pi_r(\xi)$) for the unique continuous extension to all of $L^2(M,\tau)$.

In order to state the fundamental proposition concerning $L^2(M,\tau)$ – or the standard form of the finite von Neumann algebra M, as it is referred to – we need one last bit of notation: the map $x\Omega \mapsto x^*\Omega$ is a conjugate-linear isometry from $M\Omega \subseteq L^2(M,\tau)$ onto itself. Denote its extension to $L^2(M,\tau)$ by J. It must be clear that J is an anti-unitary involution – i.e., J is a conjugate linear isometry of $L^2(M,\tau)$ onto itself whose square is the identity; in particular $J = J^* = J^{-1}$. This operator J is referred to as the *modular conjugation operator* for M, and sometimes denoted by J_M.

THEOREM 1.2.4 *(1) $JMJ = M'$.*

(2) The following conditions on a vector $\xi \in L^2(M,\tau)$ are equivalent:
(i) ξ is left-bounded;
(i)' $\xi = x\Omega$ for some (uniquely determined) element x of M;
(ii) ξ is right-bounded;
(ii)' $\xi = x'\Omega$ for some (uniquely determined) element x' of M'.

Proof: (1) The definition of J shows that if $x, y \in M$, then $(Jx^*J)(y\Omega) = yx\Omega$; since 'left multiplications commute with right multiplications', we find that

$$JMJ \subseteq M'. \tag{1.2.3}$$

On the other hand, if $x \in M, x' \in M'$, then by the 'self-adjointness' of the anti-unitary operator J, we have

$$\begin{aligned} \langle Jx'\Omega, x\Omega\rangle &= \langle Jx\Omega, x'\Omega\rangle \\ &= \langle x^*\Omega, x'\Omega\rangle \\ &= \langle \Omega, xx'\Omega\rangle \\ &= \langle \Omega, x'x\Omega\rangle \\ &= \langle x'^*\Omega, x\Omega\rangle. \end{aligned}$$

Since x was arbitrary, this implies that

$$Jx'\Omega = x'^*\Omega, \ \forall x' \in M'. \tag{1.2.4}$$

The above equation, together with the same reasoning that led to (1.2.3), now shows that

$$JM'J \subseteq M, \tag{1.2.5}$$

and the equality in (1) follows from equations (1.2.3) and (1.2.5).

(2) Denote the set of left- (resp., right-) bounded vectors in $L^2(M,\tau)$ by $\mathcal{U}_\ell$ (resp., $\mathcal{U}_r$); it is clear from the definitions that $\xi \in \mathcal{U}_\ell \Rightarrow \pi_\ell(\xi) \in (M')' = M$ and $\xi = \pi_\ell(\xi)\Omega$; conversely, if $x \in M$, then $x\Omega \in \mathcal{U}_\ell$ and $x = \pi_l(x\Omega)$. Thus the map $\xi \mapsto \pi_\ell(\xi)$ establishes a bijective linear map of $\mathcal{U}_\ell$ onto M, with inverse being given by $x \to x\Omega$. In an identical manner, we also have $\mathcal{U}_r = M'\Omega$. Thus $(i) \Leftrightarrow (i)'$ and $(ii) \Leftrightarrow (ii)'$.

To complete the proof, we only need to prove that $M\Omega = M'\Omega$. If $x \in M$, then

$$x\Omega = Jx^*J\Omega \in M'\Omega,$$

hence showing that $M\Omega \subseteq M'\Omega$. An identical reasoning, with the roles of M and M' reversed, proves the reverse inclusion, and hence the theorem. □

1.3 Discrete crossed products

The two constructions used initially by Murray and von Neumann – see [vN1] and [vN2] – to construct examples of factors of all possible types were (i) the crossed product construction, and (ii) the infinite tensor product construction. This section is devoted to a discussion of the former, while the latter will be discussed at the end of this chapter.

The starting data is a (discrete) group G acting on a von Neumann algebra M – i.e., suppose we are given a group homomorphism $t \mapsto \alpha_t$ from G to the group $\mathrm{Aut}(M)$ of (normal) *-automorphisms of M. (It is a fact (cf. [Sun1], Exercise 2.3.4(a)) that all *-algebra automorphisms of a von Neumann algebra are automatically normal.) The crossed product is a sort of maximal von Neumann algebra containing copies of M and G with 'commutation relations governed by the given action of G on M'; a little more precisely, the crossed product of M by the action α of G is a specific von Neumann algebra of the form $\tilde{M} = (\pi(M) \cup \lambda(G))''$ where $\pi : M \to \tilde{M}$ (resp., $\lambda : G \to \mathcal{U}(\tilde{M})$, the unitary group of $\tilde{M}$) is an injective normal *-homomorphism (resp., injective unitary representation) of M (resp., of G), such that the copies $\pi(M)$ and $\lambda(G)$ of M and G satisfy the commutation relations

$$\lambda(t)\pi(x)\lambda(t)^* = \pi(\alpha_t(x)), \quad \text{for all } x \in M, t \in G. \tag{1.3.1}$$

The construction of the crossed product (which is usually denoted by $M \times_\alpha G$, or simply $M \times G$) goes as follows:

Suppose $M \subseteq \mathcal{L}(\mathcal{H})$. The Hilbert space $\tilde{\mathcal{H}}$ on which $\tilde{M}$ will be represented has three (unitarily equivalent) descriptions:

(i) $\tilde{\mathcal{H}} = \ell^2(G;\mathcal{H}) = \{\xi : G \to \mathcal{H} | \sum_{t\in G} \|\xi(t)\|^2 < \infty\}$;

(ii) $\tilde{\mathcal{H}} = \bigoplus_{t\in G} \mathcal{H} = \{((\xi(t)))_{t\in G} : \sum_{t\in G} \|\xi(t)\|^2 < \infty\}$

where we think of a typical element of $\tilde{\mathcal{H}}$ as a column-vector with norm-square-summable entries from $\mathcal{H}$; although this seems an artificial variant of (i), it is this description, in view of the availability of the convenience of matrix manipulations, which will prove most useful for dealing with $M \times_\alpha G$;

(iii) $\tilde{\mathcal{H}} = \mathcal{H} \otimes \ell^2(G)$, the Hilbert space tensor product of $\mathcal{H}$ and the Hilbert space $\ell^2(G)$ of square-summable complex functions on G (or $\ell^2(G; \mathbb{C})$ in the notation of (i)).

Before proceeding further, let us remark that $\ell^2(G)$ carries two natural – the so-called left-regular and right-regular – unitary representations $\lambda, \rho : G \to \mathcal{U}(\mathcal{L}(\ell^2(G)))$, defined by $(\lambda_u \xi)(t) = \xi(u^{-1}t)$ and $(\rho_u \xi)(t) = \xi(tu)$. Alternatively, if $\{\xi_t : t \in G\}$ denotes the canonical (or distinguished) orthonormal basis of $\ell^2(G)$ – thus ξ_t is the characteristic function of the singleton set $\{t\}$ – then $\lambda_u(\xi_t) = \xi_{ut}$ and $\rho_u(\xi_t) = \xi_{tu^{-1}}$. It is a basic fact – see §1.4 – that

$$\lambda(G)' = (\rho(G))'', \rho(G)' = \lambda(G)''. \tag{1.3.2}$$

To return to the definition of the crossed product, define $\pi : M \to \mathcal{L}(\tilde{\mathcal{H}})$ and $\lambda : G \to \mathcal{L}(\tilde{\mathcal{H}})$ by:

$$\left. \begin{array}{rcl} (\pi(x)\tilde{\xi})(t) & = & \alpha_{t^{-1}}(x)\tilde{\xi}(t), \\ (\lambda(u)\tilde{\xi})(t) & = & \tilde{\xi}(u^{-1}t). \end{array} \right\} \tag{1.3.3}$$

It is trivial to verify that $\pi : M \to \mathcal{L}(\tilde{\mathcal{H}})$ (resp. $\lambda : G \to \mathcal{L}(\tilde{\mathcal{H}})$) is a faithful normal *-homomorphism (resp., faithful unitary representation) and that π and λ satisfy equation (1.2.1). Now define $\tilde{M} = (\pi(M) \cup \lambda(G))''$, so that $\tilde{M}$ is the smallest von Neumann subalgebra of $\mathcal{L}(\tilde{\mathcal{H}})$ containing $\pi(M)$ and $\lambda(G)$. This $\tilde{M}$ is, by definition, the crossed product $M \times_\alpha G$.

We used some realisation of M as a concrete von Neumann algebra to define the crossed product $M \times_\alpha G$. It is true, however – cf., for instance, [Sun1], Proposition 4.4.4 – that the isomorphism class of the von Neumann algebra $M \times_\alpha G$ so constructed does not depend upon which faithful realisation M on Hilbert space one started with.

We shall now pass to a closer analysis of $\tilde{M}$, by using the second picture of $\tilde{\mathcal{H}}$ as $\bigoplus_{t \in G} \mathcal{H}$. In this form, it is clear that any bounded operator $\tilde{x} \in \mathcal{L}(\tilde{\mathcal{H}})$ is represented by a matrix $\tilde{x} = ((\tilde{x}(s,t)))_{s,t \in G}$ where $\tilde{x}(s,t) \in \mathcal{L}(\mathcal{H})$ for all $s, t \in G$, and $(\tilde{x}\tilde{\xi})(s) = \sum_{t \in G} \tilde{x}(s,t)\tilde{\xi}(t)$, the sum on the right being intepreted as the norm limit of the net of finite sums. In this language, it is clear that

$$(\pi(x))(s,t) = \delta_{st}\alpha_{t^{-1}}(x)$$

and

$$(\lambda(u))(s,t) = \delta_{s,ut}$$

for x in M, u, s, t in G.

The point is that while elements of $\text{Mat}_G(\mathbb{C})$ have two degrees of freedom, the above generators of the crossed product have only one degree of freedom. More precisely, we have the following matricial description of the crossed product (where we identify an element $\tilde{x} \in \mathcal{L}(\tilde{\mathcal{H}})$ with its matrix $((\tilde{x}(s,t)))$ (with respect to the orthonormal basis $\{\xi_t : t \in G\}$)).

LEMMA 1.3.1 *With the preceding notation, we have*
$\tilde{M} = \{\tilde{x} \in \mathcal{L}(\tilde{\mathcal{H}}) : \exists x : G \to M \text{ s.t. } \tilde{x}(s,t) = \alpha_{t^{-1}}(x(st^{-1})) \ \forall s,t \in G\}.$

Proof: On the one hand, the right side is seen, quite easily, to define a (weakly closed self-adjoint algebra of operators and hence a) von Neumann subalgebra of $\mathcal{L}(\tilde{\mathcal{H}})$, while on the other, the algebra generated by $\pi(M) \cup \lambda(G)$ is seen to be the dense subalgebra, corresponding to finitely supported functions, of the one given by the right side. □

It follows from Lemma 1.3.1 that the crossed product $M \times_\alpha G$ is identifiable – via the association $\tilde{x} \mapsto x(s) = \tilde{x}(s,1)$ – with a space of functions from G to M – viz.

$$\tilde{M} = \{x : G \to M | \exists \tilde{x} \in \mathcal{L}(\tilde{\mathcal{H}}) \text{ s.t. } \tilde{x}(s,t) = \alpha_{t^{-1}}(x(st^{-1})) \ \forall s,t \in G\}. \quad (1.3.4)$$

It is a matter of easy verification to check that the algebra structure inherited from $(\pi(M) \cup \lambda(G))''$ by the set $\tilde{M}$ defined by equation (1.3.4) is:

$$\left.\begin{array}{rcl} (x * y)(s) & = & \sum_{t \in G} \alpha_{t^{-1}}(x(st^{-1}))y(t), \\ x^*(s) & = & \alpha_{s^{-1}}(x(s^{-1})^*). \end{array}\right\} \quad (1.3.5)$$

(The series above is interpreted as the limit, in the weak topology, of the net of finite sums; this converges by the nature of matrix multiplication.)

In the new notation, note that

$$\pi(x)(s) = \delta_{s1} \cdot x, x \in M,$$

and

$$\lambda(u)(s) = \delta_{su} \cdot 1, u \in G.$$

We conclude this section by determining when the crossed product is a finite von Neumann algebra – i.e., admits a faithful normal tracial state.

PROPOSITION 1.3.2 *Let $\tilde{M} = M \times_\alpha G$ be as above (cf. the discussion preceding equation (1.3.5)). Then $\tilde{M}$ admits a faithful normal tracial state $\tilde{\tau}$ if and only if M admits a faithful normal tracial G-invariant state τ (where G-invariance means $\tau \circ \alpha_t = \tau \ \ \forall t \in G$).*

Proof: If $\tilde{\tau}$ is a faithful normal tracial state on $\tilde{M}$, then $\tau(x) = \tilde{\tau}(\pi(x))$ defines a faithful normal tracial state on M which is G-invariant since $\tau(\alpha_t(x)) = \tilde{\tau}(\pi(\alpha_t(x))) = \tilde{\tau}(\lambda(t)\pi(x)\lambda(t)^{-1}) = \tilde{\tau}(\pi(x)) = \tau(x)$.

Conversely, if τ is a G-invariant faithful normal tracial state, define $\tilde{\tau}(x) = \tau(x(1))$ $(= \tau(\tilde{x}(1,1)))$. Then $\tilde{\tau}$ is clearly a normal positive linear function; further, $\tilde{\tau}$ is faithful (since $\tilde{x} \in \tilde{M}_+$ and $\tau(\tilde{x}(1,1)) = 0$ implies $\tilde{x}(1,1) = 0$, whence $\tilde{x}(t,t) = \alpha_{t^{-1}}(\tilde{x}(1,1)) = 0$ for all t in G, whence $\tilde{x} = 0$ (since a positive operator with zero diagonal is zero)). Finally, $\tilde{\tau}$ is a trace, since

$$
\begin{aligned}
\tilde{\tau}(x * y) &= \tau((x * y)(1)) \\
&= \tau(\sum_{t \in G} \alpha_{t^{-1}}(x(t^{-1}))y(t)) \\
&= \sum_{t \in G} \tau(\alpha_{t^{-1}}(x(t^{-1}))y(t)) \\
&\quad \text{(since } \tau \text{ is normal)} \\
&= \sum_{t \in G} \tau(x(t^{-1})\alpha_t(y(t))) \\
&\quad \text{(since } \tau \text{ is } G\text{-invariant)} \\
&= \sum_{t \in G} \tau(\alpha_t(y(t))x(t^{-1})) \\
&\quad \text{(since } \tau \text{ is a trace)} \\
&= \sum_{s \in G} \tau(\alpha_{s^{-1}}(y(s^{-1}))x(s)) \\
&= \tilde{\tau}(y * x).
\end{aligned}
$$

□

Before discussing when the crossed product is a factor, we digress for some examples, one of which will motivate the definitions of the necessary concepts.

1.4 Examples of factors

1.4.1 Group von Neumann algebras

In the notation of §1.2, the (left) group von Neumann algebra LG of the discrete group G is defined thus:

$$LG = \mathbb{C} \times G = \lambda(G)'' \subseteq \mathcal{L}(\ell^2(G)),$$

where the crossed product is with respect to the trivial action – $\alpha_t(z) = z, \ \forall t \in G, z \in \mathbb{C}$ – of G on $\mathbb{C}$, and, of course, λ denotes the left-regular representation of G on $\ell^2(G)$.

The analysis of §1.2 translates, in this most trivial case of a crossed product, as follows: the elements of LG are those $\tilde{x} \in \mathcal{L}(\ell^2(G))$ whose matrix, with respect to the standard orthonormal basis $\{\xi_t : t \in G\}$ of $\ell^2(G)$, has the form $\tilde{x}(s,t) = x(st^{-1})\ \forall s,t \in G$ for some function $x : G \to \mathbb{C}$. Note that $\tilde{x}\xi_1 = \sum_t \tilde{x}(t,1)\xi_t$, so that the function $x : G \to \mathbb{C}$ is nothing but $\tilde{x}\xi_1$. Thus there exists a unique function $\eta : LG \to \ell^2(G)$ – defined by $\eta(\tilde{x}) = \tilde{x}\xi_1$ – such that $\tilde{x}(s,t) = \eta(x)(st^{-1})\ \forall s,t$ in G.

Some of the main features of the von Neumann algebra LG are contained in the next proposition.

PROPOSITION 1.4.1 *(1) LG is a finite von Neumann algebra; more precisely, the equation $\tilde{\tau}(\tilde{x}) = \langle \tilde{x}\xi_1, \xi_1\rangle$ defines a faithful normal tracial state on LG;*

(2) $(\ell^2(G), \mathrm{id}_{LG}, \xi_1)$ is a GNS triple for the trace $\tilde{\tau}$ of (1) above, where id_{LG} is the identity representation of LG in $\ell^2(G)$; hence LG, realised as operators on $\ell^2(G)$, is in standard form in the sense of §1.2; in particular,

$$\lambda(G)' = \rho(G)'', \quad \rho(G)' = \lambda(G)''. \tag{1.4.1}$$

(3) The following conditions on G are equivalent:

(i) LG is a factor;

(ii) G is an 'ICC group' – i.e., every non-trivial conjugacy class $C(t) = \{sts^{-1} : s \in G\}$, for $t \neq 1$, is infinite.

In particular, if G is an ICC group, then LG is a II_1 factor.

Proof : (1) The equation $\tau(\lambda) = \lambda, \lambda \in \mathbb{C}$, is clearly a trace on $(M =)\ \mathbb{C}$ which is invariant under the trivial action of G and $\mathbb{C}$. Hence (the proof of) Proposition 1.3.2 implies that the equation $\tilde{\tau}(\tilde{x}) = \tilde{x}(1,1) = \eta(x)(1) = \langle \tilde{x}\xi_1, \xi_1\rangle$ defines a faithful normal tracial state on LG.

(2) $[(LG)\xi_1] \supseteq [\{\xi_t = \lambda(t)\xi_1 : t \in G\}] = \ell^2(G)$ and so ξ_1 is a cyclic vector for LG such that $\tilde{\tau}(\tilde{x}) = \langle \tilde{x}\xi_1, \xi_1\rangle\ \forall \tilde{x} \in LG$. This proves the first assertion in (2).

As for the second, note that the canonical conjugation J is the unique anti-unitary operator in $\ell^2(G)$ such that $J\xi_t = \xi_{t^{-1}}$ for all $t \in G$, and that

$$\begin{aligned} J\lambda_s J\xi_t &= J\lambda_s\xi_{t^{-1}} \\ &= J\xi_{st^{-1}} \\ &= \xi_{ts^{-1}} \\ &= \rho_s\xi_t. \end{aligned}$$

Deduce now from Theorem 1.2.4(2) that

$$\lambda(G)' = \lambda(G)''' = (LG)' = J(LG)J = J\lambda(G)''J = (J\lambda(G)J)'' = \rho(G)''. \tag{1.4.2}$$

Thus $\lambda(G)''$ and $\rho(G)''$ are commutants of one another, thus proving (2).

(3) As before, write $\eta : LG \to \ell^2(G)$ for $\eta(\tilde{x}) = \tilde{x}\xi_1$. An easy computation shows that $\tilde{x} \in Z(\tilde{M})$ if and only if $\eta(\tilde{x})$ is constant on conjugacy classes.

Suppose now that G is an ICC group. Then the conditions that $\eta(\tilde{x})$ is constant on conjugacy classes and that $\eta(\tilde{x}) \in \ell^2(G)$ force $\eta(\tilde{x})(t) = 0$ for $t \neq 1$, i.e., $\tilde{x} = (\eta(\tilde{x})(1)) \cdot \mathrm{id}_{\ell^2(G)}$. Thus $Z(LG) = \mathbb{C}$ and LG is a factor.

Conversely suppose G is not an ICC group so that there exists a finite conjugacy class $1 \neq C \subseteq G$; then if $\eta = 1_C$ denotes the characteristic function of the finite set C, it follows that $\eta \in \ell^2(G)$ and that the matrix defined by $\tilde{x}(s,t) = 1_C(st^{-1})$ defines a bounded operator on $\ell^2(G)$, so that $\tilde{x} \in LG$ and $\eta(\tilde{x}) = \eta$. The earlier discussion shows that $\tilde{x} \in Z(LG)$ (since η is constant on conjugacy classes) while $\tilde{x} \notin \mathbb{C}$, since $\eta(\tilde{x}) = \eta \notin \mathbb{C}\xi_1$; thus LG has non-trivial centre and is not a factor.

Finally, if G is an ICC group, then LG is a finite factor by (1) and (2) of this proposition; the fact that G is an infinite group shows that LG is an infinite-dimensional vector space over $\mathbb{C}$ and hence not a type I factor (since finite type I factors are (isomorphic to some $M_n(\mathbb{C})$ and hence) of finite dimension over $\mathbb{C}$). □

Examples of ICC groups are given by:

(i) $S_\infty = \bigcup_{n=1}^{\infty} S_n$ = the group of those permutations of 1,2,3,... which fix all but finitely many integers;

(ii) $\mathbf{F}_n$ $(n > 1)$, the free group on n generators.

Hence LS_∞ and $L\mathbf{F}_n, n > 1$, are (our first examples of) II_1 factors.

1.4.2 Crossed products of commutative von Neumann algebras

Let $M = L^\infty(\Omega, \mathcal{F}, \mu)$, where $(\Omega, \mathcal{F}, \mu)$ is a separable probability space. (The assumption of separability is equivalent to the requirement that $L^2(\Omega, \mathcal{F}, \mu)$ is separable. Also, it is true – see [Tak1], for instance – that such an M is the most general example of a commutative von Neumann algebra with separable pre-dual.) By an automorphism of the measure space $(\Omega, \mathcal{F}, \mu)$, we shall mean a bimeasurable bijection T of Ω such that $\mu \circ T^{-1}$ and μ have the same null sets. (Thus, $T : \Omega \to \Omega$ is a bijection satisfying: (i) if E is a subset of Ω, then $E \in \mathcal{F} \Leftrightarrow T^{-1}(E) \in \mathcal{F}$; and (ii) if $E \in \mathcal{F}$, then $\mu(E) = 0 \Leftrightarrow \mu(T^{-1}(E)) = 0$. Such a map T is also called a non-singular transformation in the literature.) The reason for the above terminology is that the most general automorphism of M is of the form $\theta(\varphi) = \varphi \circ T$, $\forall \varphi \in L^\infty(\Omega, \mathcal{F}, \mu)$, for some automorphism T of $(\Omega, \mathcal{F}, \mu)$.

If M, θ, T are as above, it is not hard to prove – cf. [Sun1], Exercise 4.1.12 – that the following conditions are equivalent:

(i) T acts 'freely' on Ω – i.e., $\mu(\{\omega \in \Omega : T\omega = \omega\}) = 0$;

(ii) $x \in M, xy = \theta(y)x \ \forall y \in M \Rightarrow x = 0$.

This is the justification for the following general definition.

DEFINITION 1.4.2 *(1) An automorphism θ of an abstract von Neumann algebra M is said to be free if $x \in M, xy = \theta(y)x \ \forall y \in M \Rightarrow x = 0$.*

(2) An action α of a group G on M – i.e., a group homormorphism $\alpha : G \to \mathrm{Aut}(M)$ – is said to be free if the automorphism α_t is free (in the sense of (1) above) for every $t \neq 1$ in G.

We need one more definition before we proceed further:

DEFINITION 1.4.3 *(1) If $\alpha : G \to \mathrm{Aut}(M)$ is an action of a group on any von Neumann algebra, write $M^\alpha = \{x \in M : \alpha_t(x) = x \ \forall t \in G\}$ for the fixed-point subalgebra of the action.*

(2) The action $\alpha : G \to \mathrm{Aut}(M)$ is said to be ergodic if $M^\alpha = \mathbb{C}\cdot 1$.

With the notation of Definition 1.4.3, it must be clear that M^α is always a von Neumann subalgebra of M. Since any von Neumann algebra is generated, as a von Neumann algebra, by its lattice of projections, it must be clear that the action α is ergodic if and only if $\mathcal{P}(M^\alpha) = \{0, 1\}$.

In the special case when $M = L^\infty(\Omega, \mathcal{F}, \mu)$, it must be clear from the foregoing discussion that any action $\alpha : G \to \mathrm{Aut}(M)$ must be of the form $\alpha_t(\varphi) = \varphi \circ T_t^{-1}$, where $t \to T_t$ is a group homomorphism from G into the group of automorphisms of the probability space $(\Omega, \mathcal{F}, \mu)$; and that the action α is ergodic if and only if

$$E \in \mathcal{F}, \mu(E \Delta T_t^{-1}(E)) = 0 \ \forall t \in G \Rightarrow \mu(E) = 0 \ \text{ or } \ \mu(E^c) = 0 \qquad (1.4.3)$$

where Δ denotes symmetric difference (so that $A \Delta B = (A - B) \cup (B - A)$) and E^c denotes the complement of the set E. (This is the classical notion of ergodicity and the reason for Definition 1.4.3(2).)

It would have been more natural to include the next proposition – which concerns general von Neumann algebras – in the last section, but it has been included here since the notions of freeness and ergodicity of an action are most natural in the context of the abelian examples considered in this section.

PROPOSITION 1.4.4 *Let $\alpha : G \to \mathrm{Aut}(M)$ be an action of a discrete group G on any von Neumann algebra M. Let $\tilde{M} = M \times_\alpha G$. Then:*

(i) $\pi(M)' \cap \tilde{M} = \pi(Z(M)) \Leftrightarrow$ the action α is free;

(ii) Assume the action α is free; then the crossed product $\tilde{M}$ is a factor if and only if the restricted action $\alpha|_{Z(M)}$ is ergodic. (Here, we use the notation $\alpha|_{M_0}$ to mean the restricted action – $(\alpha|_{M_0})_t(x_0) = \alpha_t(x_0)$ for all $x_0 \in M_0$ – of G on any von Neumann subalgebra M_0 of M which is invariant under the action α.)

In particular, if α is a free action of a discrete group G on $L^\infty(\Omega, \mathcal{F}, \mu)$, then the crossed product is a factor if and only if the action α is ergodic.

Proof: As before, if $\tilde{x}\in\tilde{M}$, we write $x=\tilde{x}\xi_1$, so that $\tilde{x}(s,t)=\alpha_{t^{-1}}(x(st^{-1}))$. In view of equation (1.3.5), it is easily seen that the condition $\tilde{x}\in\pi(M)'\cap\tilde{M}$ translates into the condition that

$$x(t)y=\alpha_{t^{-1}}(y)x(t)\ \forall y\in M,t\in G. \tag{1.4.4}$$

Hence, if the action is assumed to be free, then

$$\begin{aligned}\tilde{x}\in\pi(M)'\cap\tilde{M} &\Rightarrow x(t)=0\ \forall t\neq 1\\ &\Rightarrow \tilde{x}=\pi(x(1))\in\pi(M).\end{aligned}$$

Since the map $\pi:M\to\tilde{M}$ is injective, we see that indeed $\tilde{x}\in\pi(M)'\cap\tilde{M}\Rightarrow\tilde{x}=\pi(z)$ for some $z(=x(1))\in Z(M)$.

Conversely, if the action is not free, there exist $t\neq 1$ in G and $x_0\in M$ such that $x_0\neq 0$ and $x_0y=\alpha_t(y)x_0$ for all $y\in M$. If we now define the function $x:G\to M$ by $x(s)=\delta_{st^{-1}}x_0$, it is clear that the equation $\tilde{x}(s,u)=\alpha_{u^{-1}}(x(su^{-1}))$ defines a bounded operator $\tilde{x}$ on $\tilde{\mathcal{H}}$; further, it must be equally clear that $\tilde{x}\in\pi(M)'\cap\tilde{M}$ and that $\tilde{x}\notin\pi(M)$ (since $((\tilde{x}(s,u)))$ is not a diagonal matrix).

(ii) Assume the action α is free. It then follows from (i) above that

$$\begin{aligned}Z(\tilde{M}) &= \pi(Z(M))\cap\lambda(G)'\\ &= \{\pi(z):z\in Z(M)\ \text{ and }\ z=\alpha_t(z)\ \forall t\in G\},\end{aligned}$$

whence the desired conclusion. □

Thus if a countable group G acts freely and ergodically on $M=L^\infty(\Omega,\mathcal{F},\mu)$, then $\tilde{M}=M\times_\alpha G$ is a factor. The complete details of the Murray–von Neumann type of the factor thus obtained are contained in the following beautiful theorem due to von Neumann ([vN1]).

THEOREM 1.4.5 *Let $\tilde{M}=M\times_\alpha G$ be as above. (Thus, $M=L^\infty(\Omega,\mathcal{F},\mu)$, where $(\Omega,\mathcal{F},\mu)$ is a separable probability space, and G is assumed to act freely and ergodically on M via the equations $\alpha_t(\varphi)=\varphi\circ T_t^{-1}$, where $\{T_t:t\in G\}$ is a group of automorphisms of the measure space $(\Omega,\mathcal{F},\mu)$.)*

(1) $\tilde{M}$ is of type I if and only if $(\Omega,\mathcal{F},\mu)$ contains atoms – i.e., $\exists E\in\mathcal{F}$ such that $\mu(E)>0$ and whenever $E_0\in\mathcal{F},E_0\subseteq E$, either $\mu(E_0)=0$ or $\mu(E-E_0)=0$. In this case, there exists a countable partition $\Omega=\coprod_{i\in\Lambda}E_i$, where each E_i is an atom; further, M is of type $I_n, 1\leq n\leq\infty$, if and only if $|\Lambda|=n$.

(2) M is of type II if and only if there exists a σ-finite measure ν on $(\Omega,\mathcal{F})$ such that (i) ν is non-atomic – i.e ν has no atoms, (ii) ν and μ have the same null sets, and (iii) ν is G-invariant – i.e., $\nu\circ T_t^{-1}=\nu\ \forall t\in G$. Further, $\tilde{M}$ is of type II_1 or II_∞ according as $\nu(\Omega)<\infty$ or $\nu(\Omega)=\infty$.

(3) $\tilde{M}$ is of type III if and only if there exists no σ-finite G-invariant measure ν which is mutually absolutely continuous with μ.

We shall not prove the theorem here. (The interested reader may find such a proof, for instance, in [Sun1], Theorem 4.3.13.) We shall, instead, content ourselves with a discussion of the II_1 case, which is the case of interest in this book.

We know from Proposition 1.3.2 that $\tilde{M}$ is a finite factor if and only if there exists a faithful normal tracial state τ on $M = L^\infty(\Omega, \mathcal{F}, \mu)$ (equivalently, if there exists a probability measure ν which has the same null sets as μ) which is G-invariant. If ν is non-atomic, then M – and hence $\tilde{M}$ – is infinite-dimensional, so that $\tilde{M}$ cannot be a factor of type $I_n, n < \infty$. Since $\tilde{M}$ is a finite factor, the non-atomicity of ν implies that $\tilde{M}$ is a factor of type II_1. On the other hand, if ν is atomic, the finiteness of ν and the ergodicity of the action imply that Ω is the union of a finite number, say n, of atoms. This means that $M \cong \mathbb{C}^n$. The assumed freeness (and ergodicity) of the action now forces G to have exactly n elements, and consequently $\tilde{M}$ is a finite factor with n^2 elements – i.e., it is a I_n factor.

EXAMPLE 1.4.6 *Suppose G is a countably infinite dense subgroup of a compact group.*

(Two examples to bear in mind are:

(a) $K = \mathbb{T} = \{e^{i\theta} : \theta \in \mathbb{R}\}, G = \{e^{i2n\pi\theta_0} : n \in \mathbb{Z}\}$, *where* θ_0 *is irrational; and*

(b) $K = \{0,1\}^{\mathbb{N}}$ = *set of all* $(0,1)$ *sequences, with group operation being coordinate-wise addition* mod 2, *and* G = *the subgroup of those sequences which contain at most finitely many 1's.)*

Set $\Omega = K$, $\mathcal{F}$ = *the* σ*-algebra of Borel sets in* K, *and* μ = *Haar measure. (In example (a),* μ *is normalised arc-length, so that* $\mu(I) = \theta/2\pi$ *if* I *is an arc subtending an angle* θ *at the origin, and in example (b),* μ *is the countable product of the measure on* $\{0,1\}$ *which assigns* $1/2$ *to* $\{0\}$ *and to* $\{1\}$.*)*

For $t \in G$ *and* $k \in K$, *define* $T_t(k) = tk$. *It should be clear that* $t \to T_t$ *is a homomorphism from* G *into the group of automorphisms of the measure space* $(K, \mathcal{F}, \mu)$. *Further, if we define* $\alpha_t(\phi) = \phi \circ T_{t^{-1}}, t \in G, \phi \in L^\infty(K, \mathcal{F}, \mu)$, *it follows that* α *is an action of* G *on* $L^\infty(K, \mathcal{F}, \mu)$ *which preserves* μ *and is ergodic (since* G *is dense in* K*).*

Hence $L^\infty(K) \times_\alpha G$ *is a* II_1 *factor, whenever* G *is an infinite countable dense subgroup of a compact group.*

EXAMPLE 1.4.7 *A slight variation of the construction in the preceding example yields factors of type III – the so-called* Powers factors *– see [Pow]. Suppose* $\Omega = \{0,1\}^{\mathbb{N}}$, *equipped with the Borel* σ*-algebra. Now take* μ_λ *to be the countable product of a measure* μ_0 *on* $\{0,1\}$ *which assigns unequal masses to* $\{0\}$ *and* $\{1\}$ *thus:* $\mu_0\{0\} = 1/(1+\lambda)$ *and* $\mu_0\{1\} = \lambda/(1+\lambda)$, *where* $\lambda \neq 1$. *If* $T_t, t \in G$, *is defined as before, it is true that each* T_t *is still an automorphism* *of* $(\Omega, \mathcal{F}, \mu)$, *although* $\mu \circ T_t^{-1} \neq \mu$ *for* $t \neq 1$. *It is still the case that* α, *as defined before, is an ergodic action of* G *on* $L^\infty(\Omega, \mathcal{F}, \mu)$. *The crucial difference is that, now, the action does not admit any* σ*-finite invariant measure which is mutually absolutely continuous with* μ, *and hence* $R_\lambda = L^\infty(\Omega, \mathcal{F}, \mu_\lambda) \times_\alpha G$ *is a factor of type III. In fact, it turns out that* R_λ *and* $R_{\lambda'}$ *are non-isomorphic for* $0 < \lambda \neq \lambda' < 1$. *We shall not prove these facts; the interested reader may consult [Sun1], for instance.*

EXAMPLE 1.4.8 *The group* $G = SL(2, \mathbb{Z})$ *admits a natural (linear) action on* $(\mathbb{R}^2, \mathcal{B}, \mu)$ *– where* μ *is Lebesgue measure defined on the* σ*-algebra* $\mathcal{B}$ *of Borel sets in* $\mathbb{R}^2$. *It is true – cf. [Zim], Example 2.2.9 – that this action is ergodic. Since* μ *is an infinite measure, it follows from Theorem 1.4.5 that* $\tilde{M} = L^\infty(\mathbb{R}^2, \mathcal{B}, \mu) \times SL(2, \mathbb{Z})$ *is a factor of type* II_∞. *(Note that Lebesgue measure is preserved by linear transformations of determinant 1.)*

In the next proposition, we single out a certain feature of the preceding example that we shall need later.

PROPOSITION 1.4.9 *Let* $\tilde{M} = M \times_\alpha G$, *where* $M = L^\infty(\mathbb{R}^2, \mu)$ *and* $G = SL(2, \mathbb{Z})$, *are as in Example 1.4.8. Let* $p_1 = 1_{\mathbb{D}}$ *denote the characteristic function of the unit disc* $\mathbb{D} = \{(x, y) \in \mathbb{R}^2 : x^2 + y^2 < 1\}$ *in* $\mathbb{R}^2$ *(so that* p_1 *is a projection in* M*). Then*

(a) $R = \pi(p_1)\tilde{M}\pi(p_1)$ *is a* II_1 *factor, where* $\pi : M \to \tilde{M}$ *is the canonical embedding; and*

(b) $R \cong pRp$ *for any non-zero projection* p *in* R.

Proof : We use the notation of §1.3, so that a typical element of $\tilde{M}$ is represented by a matrix of the form $((\tilde{x}(s, t)))_{s,t \in G}$ – which induces a bounded operator on $\bigoplus_{t \in G} \mathcal{H}_t, \mathcal{H}_t = L^2(\mathbb{R}^2, \mu) \; \forall t \in G$ – such that $\tilde{x}(s, t) = \alpha_{t^{-1}}(x(st^{-1}))$, where $\{x(s) : s \in G\} \subseteq L^\infty(\mathbb{R}^2, \mu)$.

Since $\tilde{M}$ is a factor and $\pi(p_1) \in \mathcal{P}(\tilde{M})$, it follows that $R = \tilde{M}_{\pi(p_1)} = \pi(p_1)\tilde{M}\pi(p_1)$ is a factor. Further, the non-atomicity of μ shows that M_{p_1}, and hence R, is infinite-dimensional over $\mathbb{C}$. To show that R is a II_1 factor, it suffies, therefore, to exhibit a faithful normal tracial state on R.

More generally, let $p_r = 1_{\mathbb{D}_r}$, where $\mathbb{D}_r = \{(x, y) \in \mathbb{R}^2 : x^2 + y^2 < r^2\}$, for $0 < r < \infty$. It is easy to see that if $\tilde{x} \in \tilde{M}$, then $\tilde{x} \in \tilde{M}_{\pi(p_r)}$, if and only if

$$\tilde{x}(s, t) = \alpha_{s^{-1}}(p_r)\tilde{x}(s, t)\alpha_{t^{-1}}(p_r); \tag{1.4.5}$$

in other words, $\tilde{x} \in \tilde{M}_{\pi(p_r)}$ if and only if $\tilde{x}(s, t)$ (which is an L^∞-function on $\mathbb{R}^2$) is supported in $T_t^{-1}(\mathbb{D}_r) \cap T_s^{-1}(\mathbb{D}_r)$. In particular, if $\tilde{x} \in \tilde{M}_{\pi(p_r)}$, then $\tilde{x}(1, 1) \in L^\infty(\mathbb{D}_r, \mu)$. It is not hard, using the fact that $\mu \circ T_t^{-1} = \mu \; \forall t \in G$, to show that the equation

$$\tau_r(\tilde{x}) = \frac{1}{\mu(\mathbb{D}_r)} \int \tilde{x}(1, 1) d\mu \tag{1.4.6}$$

defines a faithful normal tracial state on $\tilde{M}_{\pi(p_r)}$. Hence $\tilde{M}_{\pi(p_r)}$ is a II_1 factor for all $r > 0$, thus proving (a).

Next, begin by noting that if $\lambda > 0$, then the equation $(\theta_\lambda(f)(x) = f(\lambda x), x \in \mathbb{R}^2, f \in L^\infty(\mathbb{R}^2, \mu)$, defines, for each $\lambda > 0$, an automorphism θ_λ of $L^\infty(\mathbb{R}, \mu) = M$. Since linear maps commute with scalar multiplication, it must be clear that $\theta_\lambda \circ \alpha_t = \alpha_t \circ \theta_\lambda \; \forall \lambda > 0, t \in SL(2, \mathbb{Z})$. This implies that there exists, for each $\lambda > 0$, a unique automorphism $\tilde{\theta}_\lambda$ of $\tilde{M}$ such that $\tilde{\theta}_\lambda((\tilde{x}(s,t))) = ((\theta_\lambda(\tilde{x}(s,t))))$.

Since $\theta_\lambda(1_{\mathbb{D}_r}) = 1_{\mathbb{D}_{\lambda r}}$ for $r, \lambda > 0$, it follows from equation (1.4.5) that $\tilde{\theta}_\lambda(\tilde{M}_{\pi(p_r)}) = \tilde{M}_{\pi(p_{\lambda r})}$; in particular, setting $r = 1$ and $0 < \lambda < 1$ we find (since $\tilde{M}_{\pi(p_\lambda)} = \tilde{M}_{\pi(p_1 p_\lambda)} = (\tilde{M}_{\pi(p_1)})_{\pi(p_\lambda)} = R_{\pi(p_\lambda)})$ that $R \cong pRp$ if $p = \pi(p_\lambda), 0 < \lambda < 1$.

Note, finally, that, if $0 < \lambda < 1$, then $\tau_1(\pi(p_\lambda)) = \frac{1}{\pi}\mu(\mathbb{D}_\lambda) = \lambda^2$, where τ_1 is as defined in equation (1.4.6), with $r = 1$). Hence if $0 \neq p \in \mathcal{P}(R)$ and $\lambda = \tau_1(p)^{1/2}$, then $\tau_1(p) = \tau_1(\pi(p_\lambda))$; so there exists (by equation (1.1.1)) a unitary $u \in R$ such that $upu^* = \pi(p_\lambda)$, and consequently $R_p \cong R_{\pi(p_\lambda)}$ (via the isomorphism $R_p \ni x \to uxu^*$), thus completing the proof of the proposition. □

1.4.3 Infinite tensor products

Fix an integer N, and let

$$A_n = \bigotimes^n M_N(\mathbb{C}) = M_{N^n}(\mathbb{C}).$$

Regard A_n as a subalgebra of A_{n+1} via the embedding

$$x \mapsto \begin{bmatrix} x & 0 & 0 & \cdots & 0 \\ 0 & x & 0 & \cdots & 0 \\ 0 & 0 & x & \cdots & 0 \\ \vdots & \vdots & \vdots & \ddots & \vdots \\ 0 & 0 & 0 & \cdots & x \end{bmatrix} = x \otimes 1_N.$$

Then it is clear that $A_\infty = \bigcup A_n$ is a *-algebra, and that there exists a unique tracial state on A_∞ which restricts on the I_{N^n} factor A_n to the unique tracial state tr_{A_n}. (In fact, it is clear that this is the only tracial state on A_∞ since finite factors admit unique tracial states.) Let $(\mathcal{H}, \pi, \Omega)$ denote the associated GNS triple. We define

$$R_{(N)} = \bigotimes^{\mathbb{N}} M_N(\mathbb{C}) = \pi(A_\infty)''. \tag{1.4.7}$$

The faithfulness of tr_{A_n} implies that π embeds A_∞ as a weakly dense *-subalgebra of $R_{(N)}$. Since the equation

$$\mathrm{tr}_{R_{(N)}}(x) = \langle x\Omega, \Omega \rangle$$

clearly defines a tracial state on $R_{(N)}$, it follows that the von Neumann algebra $R_{(N)}$ has the property of admitting a unique normal tracial state. This implies that $R_{(N)}$ is a II_1 factor. (Reason: if $h \in Z(R_{(N)})_+$ is arbitrary, then also $\mathrm{tr}_{R_{(N)}}(h\cdot)$ is a trace; further, in view of the infinite-dimensionality of A_∞, the finite factor $R_{(N)}$ cannot be of type I.)

(Here is an alternative way of seeing that $R_{(N)}$ is a II_1 factor: consider the compact group $K = \mathbb{Z}_N^{\mathbb{N}}$ obtained as the direct product of countably infinitely many copies of the discrete group $\mathbb{Z}_N$; let G be the dense subgroup consisting of those sequences from $\mathbb{Z}_N$ which differ from the identity element in at most finitely many places; then $R_{(N)}$ is isomorphic to the II_1 factor constructed as in Example 1.4.6.)

Notice that $R_{(N)}$ is clearly approximately finite-dimensional in the sense of the following definition. Not quite so clear, however, is the striking statement of the subsequent theorem due to Murray and von Neumann (which we shall not prove here). (See [MvN3] and also [Con1].)

DEFINITION 1.4.10 *A von Neumann algebra M is said to be approximately finite dimensional, or simply AFD, if there exists an increasing sequence*

$$A_1 \subseteq A_2 \subseteq \cdots \subseteq A_n \subseteq A_{n+1} \subseteq \cdots$$

*of finite-dimensional *-subalgebras, whose union in weakly dense in M.*

THEOREM 1.4.11 *Every AFD II_1 factor is isomorphic to $R_{(2)}$.*

Thus we find that among the class of all II_1 factors, there is one, the so-called *hyperfinite II_1 factor* – which we shall always denote by the symbol R – which is uniquely determined up to isomorphism by the property that it is approximately finite-dimensional. Thus, for instance, if $S_\infty = \bigcup_{n=1}^\infty S_n$ is the group of those permutations of $\mathbb{N}$ which move at most finitely many integers, then the group von Neumann algebra LS_∞ is one model of the hyperfinite II_1 factor. Less obvious, but also true, is the fact that the algebra denoted by R in Proposition 1.4.9 is another model for the hyperfinite II_1 factor.

Chapter 2

Subfactors and index

2.1 The classification of modules

If M is any von Neumann algebra, we shall, by an *M-module*, mean a Hilbert space $\mathcal{H}$ equipped with an action of M, i.e., a unital normal homomorphism from M into $\mathcal{L}(\mathcal{H})$. We shall generally suppress specific mention of the underlying representation $\pi : M \to \mathcal{L}(\mathcal{H})$ and just write $x\xi$ instead of $\pi(x)\xi$ when $x \in M, \xi \in \mathcal{H}$.

If $\mathcal{H}$ and $\mathcal{K}$ are M-modules, we shall say that a bounded linear operator $T : \mathcal{H} \to \mathcal{K}$ is *M-linear* if $T(x\xi) = x(T\xi)\ \forall x \in M, \xi \in \mathcal{H}$, and we shall denote the collection of all such M-linear operators from $\mathcal{H}$ to $\mathcal{K}$ by ${}_M\mathcal{L}(\mathcal{H}, \mathcal{K})$. When $\mathcal{H} = \mathcal{K}$, we shall simply write ${}_M\mathcal{L}(\mathcal{H})$ for ${}_M\mathcal{L}(\mathcal{H}, \mathcal{H})$. (If π denotes the underlying representation of M on $\mathcal{H}$, note that ${}_M\mathcal{L}(\mathcal{H}) = \pi(M)'$ and consequently ${}_M\mathcal{L}(\mathcal{H})$ is a von Neumann algebra. In general, it is true similarly that ${}_M\mathcal{L}(\mathcal{H}, \mathcal{K})$ is a weakly closed subspace of $\mathcal{L}(\mathcal{H}, \mathcal{K})$ which is well-behaved under polar decomposition in the sense that if $T \in \mathcal{L}(\mathcal{H}, \mathcal{K})$ has polar decomposition $T = U|T|$, then T is M-linear if and only if U and $|T|$ are.)

Two M-modules $\mathcal{H}$ and $\mathcal{K}$ are said to be *isomorphic* if there exists an M-linear unitary operator of $\mathcal{H}$ onto $\mathcal{K}$. (In view of the parenthetical general remark of the last paragraph, this is equivalent to the existence of an invertible M-linear map of $\mathcal{H}$ onto $\mathcal{K}$.)

EXAMPLE 2.1.1 *(i) If ϕ is a normal state on any von Neumann algebra M, we shall use – here and elsewhere in the sequel – the symbol $L^2(M, \phi)$ to denote the Hilbert space underlying the GNS representation of M associated with ϕ, and we shall denote the cyclic vector by Ω_ϕ. Then, by definition, the Hilbert space $L^2(M, \phi)$ is an M-module. (Further, as has already been remarked, every cyclic M-module is isomorphic to such a module, for some normal state ϕ.)*

(ii) If $\{\mathcal{H}_i : i \in I\}$ is any family of M-modules, then their direct sum $\bigoplus_{i \in I} \mathcal{H}_i$ is again an M-module. In particular, if $I = \mathbb{N}$ and $\mathcal{H}_i \cong \mathcal{H}\ \forall i$, this direct sum will be denoted by $\mathcal{H} \otimes \ell^2$ (for the obvious reason that the

underlying Hilbert spaces are isomorphic and the action, in the tensor-product formulation, can be identified thus: $x(\xi \otimes \eta) = \pi(x)\xi \otimes \eta)$.

The following result is crucial in the classification of modules over factors.

PROPOSITION 2.1.2 *Let M be any von Neumann algebra and let $\mathcal{K}$ be a faithful M-module – i.e., a module such that the underlying representation of M is faithful . Then any separable M-module $\mathcal{H}$ is isomorphic to a submodule of $\mathcal{K} \otimes \ell^2$.*

Proof: First consider the case when $\mathcal{H}$ is a cyclic M-module and consequently isomorphic to $L^2(M, \psi)$ for some normal state ψ on M. The faithfulness assumption says that M is embedded as a von Neumann subalgebra of $\mathcal{L}(\mathcal{K})$ and consequently, we may (extend ψ to a normal state on all of $\mathcal{L}(\mathcal{K})$ – thanks to the Hahn–Banach extension theorem – and consequently) find a sequence $\{\xi_n\}$ of vectors in $\mathcal{K}$ such that $\sum_n ||\xi_n||^2 = 1$ and $\psi(x) = \sum_n \langle x\xi_n, \xi_n \rangle \; \forall x \in M$.

Notice now that, for each fixed n, we have, for all $x \in M$,

$$||x\xi_n||^2 = \langle x^*x\xi_n, \xi_n \rangle \leq \psi(x^*x) = ||x\Omega_\psi||^2; \tag{2.1.1}$$

hence there exists a unique bounded operator $R_n : L^2(M, \psi) \rightarrow \mathcal{K}$ such that $R_n(x\Omega_\psi) = x\xi_n \; \forall x \in M$. It follows at once from the definitions that each R_n is actually an M-linear map.

Now consider the (clearly M-linear) operator $U : L^2(M, \psi) \rightarrow \mathcal{K} \otimes \ell^2$ defined by $U\xi = \sum_n R_n\xi \otimes \epsilon_n$ where $\{\epsilon_n\}_{n=1}^\infty$ denotes the standard orthonormal basis for ℓ^2. It must be clear that U is an isometric operator, whence $L^2(M, \psi)$ – and consequently $\mathcal{H}$ – is isomorphic, as an M-module, to a submodule (viz., the range of the operator U) of $\mathcal{K} \otimes \ell^2$.

Thus we have proved the proposition for cyclic modules. The general case follows from the facts that (i) every separable module is isomorphic to a countable direct sum of cyclic modules, and (ii) as an M-module, a countable direct sum of copies of the module $\mathcal{K} \otimes \ell^2$ is isomorphic to $\mathcal{K} \otimes \ell^2$ itself. □

REMARK 2.1.3 *One consequence of the preceding result is a fact which is sometimes termed the 'structure of normal isomorphisms'. Suppose then that $M_i \subseteq \mathcal{L}(\mathcal{H}_i), i = 1, 2$, are von Neumann algebras and suppose that there exists a normal isomorphism θ of M_1 onto M_2. Then put $M = M_1$ and regard θ as a faithful representation of M, and deduce from Proposition 2.1.2 that $\mathcal{H}_1$ is isomorphic, as an M-module, to a submodule of $\mathcal{H}_2 \otimes \ell^2$. Similarly, $\mathcal{H}_2$ is isomorphic, as an M-module, to $\mathcal{H}_1 \otimes \ell^2$.*

Also, it follows that each of the M-modules, $\mathcal{H}_i \otimes \ell^2, i = 1, 2$, is isomorphic to a submodule of the other. This implies – by the same reasoning as is employed to prove the classical Schroeder–Bernstein theorem – that these two modules are actually isomorphic. We thus have the so-called 'structure *of normal isomorphisms', viz.: the isomorphism θ is a composition of three maps, (i) a dilation or 'ampliation' (* $\mathrm{id} \otimes 1_{\ell^2}$*), (ii) a spatial isomorphism – i.e., one implemented by a unitary isomorphism of the underlying Hilbert spaces (the one establishing the isomorphism of the M-modules $\mathcal{H}_i \otimes \ell^2$), and (iii) a 'reduction' – i.e., the restriction to a submodule.*

For us, the importance of Proposition 2.1.2 lies in the classification of modules. For one thing, it says that it is enough to classify, up to isomorphism, the various submodules of one module – which we may choose as $\mathcal{H}_\infty = \mathcal{H}_1 \otimes \ell^2$, where $\mathcal{H}_1 = L^2(M, \phi)$ for some arbitrary faithful normal state ϕ on M.

Before proceeding further, we single out two simple facts as lemmas, for ease of reference.

LEMMA 2.1.4 *Let M be a von Neumann algebra and let $\mathcal{K}$ be an M-module. Then the map $p \mapsto \mathrm{ran}\, p$ sets up a bijection between the set $\mathcal{P}({}_M\mathcal{L}(\mathcal{K}))$ of M-linear projection operators in $\mathcal{K}$ and the set of M-submodules of $\mathcal{K}$. Further, two projections $p, q \in \mathcal{P}({}_M\mathcal{L}(\mathcal{K}))$ are Murray–von Neumann equivalent (relative to the von Neumann algebra ${}_M\mathcal{L}(\mathcal{K})$) if and only if their ranges are isomorphic as M-modules.*

Proof: Easy. □

As in our treatment of crossed products, we adopt the convention that if $\mathcal{K}$ is a Hilbert space, then $\mathcal{K} \otimes \ell^2$ is identified with the Hilbert space direct sum of countably infinitely many copies of $\mathcal{K}$; also, we shall identify an operator $T \in \mathcal{L}(\mathcal{K} \otimes \ell^2)$ with an infinite matrix $((T_{ij}))$ with entries from $\mathcal{L}(\mathcal{K})$ (in such a way that if $x \in \mathcal{L}(\mathcal{K})$, then the operator $x \otimes \mathrm{id}_{\ell^2}$ gets identified with the matrix with x on each diagonal entry and zeros elsewhere).

Recall that if $M \subseteq \mathcal{L}(\mathcal{K}), N \subseteq \mathcal{L}(\ell^2)$ are von Neumann algebras, then $M \otimes N$ denotes the von Neumann subalgebra of $\mathcal{K} \otimes \ell^2$ generated by operators of the form $x \otimes y, x \in M, y \in N$. Under the identification discussed in the previous paragraph, it should be fairly clear that $M \otimes \mathcal{L}(\ell^2)$ corresponds to the set of those operators T on $\mathcal{K} \otimes \ell^2$ for which all the entries of the associated matrix $((T_{ij}))$ come from the von Neumann algebra M.

LEMMA 2.1.5 *If $M \subseteq \mathcal{L}(\mathcal{K})$ is a von Neumann algebra, then*

$$(M \otimes 1)' = M' \otimes \mathcal{L}(\ell^2). \tag{2.1.2}$$

Proof: This is a routine computation. □

Now we are ready for the classification of separable modules over factors with separable pre-duals.

THEOREM 2.1.6 *Let M be a factor with separable pre-dual and let $\mathcal{H}$ be a separable M-module.*

(i) If M is of type I, there exists a sequence $\{\mathcal{H}_n : n \in \overline{\mathbb{N}} = \{1, 2, \cdots, \infty\}\}$ of pairwise non-isomorphic M-modules, and there exists (a necessarily unique) $n \in \overline{\mathbb{N}}$ such that $\mathcal{H} \cong \mathcal{H}_n$.

(ii) If M is of type II, there exists a family $\{\mathcal{H}_d : d \in \overline{\mathbb{R}} = [0, \infty]\}$ of pairwise non-isomorphic M-modules, and there exists (a necessarily unique) $d \in \overline{\mathbb{R}}$ such that $\mathcal{H} \cong \mathcal{H}_d$.

(iii) If M is of type III, there exists a separable non-zero M-module which is unique up to isomorphism.

Proof: We consider the several possible cases.

Case(i): M is of type I As mentioned in §1.1, there exists a separable Hilbert space $\mathcal{K}$ such that $M \cong \mathcal{L}(\mathcal{K})$. Deduce from Lemma 2.1.5 that $\tilde{M} = {}_M\mathcal{L}(\mathcal{K} \otimes \ell^2)$ $(= 1 \otimes \mathcal{L}(\ell^2))$ is a factor of type I_∞; it follows that for any projection $p \in \tilde{M}$, it is the case that $\tilde{M}_p = p\tilde{M}p$ is again a factor of type I. On the other hand, it is a consequence of Proposition 2.1.2 that if $\mathcal{H}$ is an arbitrary M-module, then $\mathcal{H}$ is isomorphic, as an M-module, to $p(\mathcal{K} \otimes \ell^2)$ for some $p \in \tilde{M}$ and consequently, that ${}_M\mathcal{L}(\mathcal{H})$ is a factor of type I and that in fact, $\mathcal{H}$ is isomorphic, as an M-module, to the direct sum of n copies of $\mathcal{K}$, if $p = 1 \otimes p_0, p_0 \in \mathcal{L}(\ell^2)$ and $n = \dim \operatorname{ran} p_0$.

Case(ii)$_1$: M is of type II_1 In this case, set $\mathcal{K} = L^2(M, \operatorname{tr})$, where, of course, the symbol tr denotes the unique normal tracial state on M. It follows from Theorem 1.2.4 and Lemma 2.1.5 that ${}_M\mathcal{L}(\mathcal{K}) = JMJ$, and consequently, ${}_M\mathcal{L}(\mathcal{K} \otimes \ell^2) = JMJ \otimes \mathcal{L}(\ell^2)$ is a factor of type II_∞.

On the other hand, it follows from Proposition 2.1.2 and Lemma 2.1.4 that the set of isomorphism classes of M-modules is in bijective correspondence with the set of Murray–von Neumann equivalence classes of projections in this II_∞ factor; the latter set is, according to the discussion in §1.1, in bijection with $[0, \infty]$.

It follows that if $\mathcal{H}$ is any separable M-module, then there exists $p \in \mathcal{P}({}_M\mathcal{L}(\mathcal{K} \otimes \ell^2))$ such that M is isomorphic, as an M-module, to $\operatorname{ran} p$ and that ${}_M\mathcal{L}(\mathcal{H})$ is a factor of type II; further, the isomorphism class of $\mathcal{H}$ depends only upon $\operatorname{Tr} p$.

Case(ii)$_\infty$: M is of type II_∞ In this case – as already mentioned in §1.1 – there exists a finite projection $p_1 \in M$ such that $M_0 = p_1 M p_1$ is a factor of type II_1, and such that $M \cong M_0 \otimes \mathcal{L}(\ell^2)$. Set $\mathcal{K}_0 = L^2(M_0, \operatorname{tr})$ and $\mathcal{K} = \mathcal{K} \otimes \ell^2$. It is seen then that ${}_M\mathcal{L}(\mathcal{K}) = J_{M_0} M_0 J_{M_0} \otimes 1$ – where, of course, the symbol J_{M_0} denotes the 'modular conjugation' operator on $\mathcal{K}_0$ – which is a II_1 factor. It follows, as in the last case, that if $\mathcal{H}$ is any separable M-module, then there exists a projection $p \in \mathcal{P}({}_M\mathcal{L}(\mathcal{K} \otimes \ell^2))$ such that M is isomorphic, as an M-module, to $\operatorname{ran} p$ and that ${}_M\mathcal{L}(\mathcal{H})$ is a factor of type II .

Case (iii): M is of type III Notice from the cases discussed so far, that if $P \subseteq \mathcal{L}(\mathcal{M}), \mathcal{M}$ an arbitrary Hilbert space, and if P is a factor of type I (resp., II), then also P' is a factor of type I (resp., II). Hence, if P is a factor of type III, it follows by exclusion that so also is P'.

Suppose now that $\mathcal{K}$ is an arbitrary faithful M-module. It follows from the previous paragraph that ${}_M\mathcal{L}(\mathcal{K} \otimes \ell^2)$ is a factor of type III. Since any two non-zero projections in such a factor are Murray–von Neumann equivalent, the proof of this case follows at once from Proposition 2.1.2 and Lemma 2.1.4. □

2.2 $\dim_M \mathcal{H}$

In order to really be able to use Theorem 2.1.6, we shall find it convenient to work with bimodules. For this reason, we shall make a slight change in terminology: thus, if $\mathcal{H}$ is what we have so far been calling an M-module, we shall henceforth refer to $\mathcal{H}$ as a *left M-module.*

On the other hand, a Hilbert space $\mathcal{H}$ is called a *right M-module* – where M is an arbitrary von Neumann algebra – if there exists a σ-weakly continuous linear map $\pi_r : M \to \mathcal{L}(\mathcal{H})$ which preserves adjoints and reverses products (i.e., $\pi_r(x^*) = (\pi_r(x))^*$ and $\pi_r(xy) = \pi_r(y)\pi_r(x)$ for all $x, y \in M$). As with left modules, we shall often omit referring to the map π_r, and simply write ξx instead of $\pi_r(x)\xi$ whenever $x \in M, \xi \in \mathcal{H}$. (The assumed product-reversal is consistent with writing the operator on the right, as above, in the sense that $\xi(xy) = (\xi x)y \; \forall x, y \in M, \xi \in \mathcal{H}$.)

Every statement about left modules has a corresponding statement about right modules, via the following observation. If M is a von Neumann algebra, recall that there is an *opposite* von Neumann algebra M^{op} such that there exists a linear isometry $x^0 \mapsto x$ from M^{op} onto M which preserves adjoints and reverses products. (Note that the conditions of the previous sentence ensure that $(M^{op})_* \cong M_*$ and determine the von Neumann algebra M^{op} uniquely up to isomorphism.) Thus, we may – and shall – think of a right M-module as a left M^{op}-module, i.e., as a Hilbert space $\mathcal{H}$ equipped with a unital normal homomorphism $\pi^{op} : M^{op} \to \mathcal{L}(\mathcal{H})$ – so that $\pi^{op}(x^0)\xi = \xi x \; \forall x \in M, \xi \in \mathcal{H}$.

DEFINITION 2.2.1 *(i) If M, N are von Neumann algebras, a Hilbert space $\mathcal{H}$ is said to be an M-N-bimodule if:*

(a) $\mathcal{H}$ is a left M-module;

(b) $\mathcal{H}$ is a right N-module; and

(c) the actions of M and N commute; i.e., $(m\xi)n = m(\xi n) \; \forall m \in M, n \in N, \xi \in \mathcal{H}$.

Thus, in order for $\mathcal{H}$ to be an M-N-bimodule, there must exist unital normal homomorphisms $\pi_l : M \to \mathcal{L}(\mathcal{H})$ and $\pi_r^0 : N^{op} \to \mathcal{L}(\mathcal{H})$ (i.e., a unital normal anti-homomorphism $\pi_r : N \to \mathcal{L}(\mathcal{H})$) such that $\pi_r^0(N^{op})$ $(= \pi_r(N)) \subseteq \pi_l(M)'$.)

(ii) If $\mathcal{H}, \mathcal{K}$ are M-N-bimodules, an operator $T \in \mathcal{L}(\mathcal{H}, \mathcal{K})$ will be called M-linear (resp., N-linear) if $T(x\xi) = xT\xi$ (resp., $T(\xi y) = (T\xi)y$) for all $x \in M, y \in N, \xi \in \mathcal{H}$. The collection of such M-linear (resp., N-linear) operators will be denoted by ${}_M\mathcal{L}(\mathcal{H}, \mathcal{K})$ (resp., $\mathcal{L}_N(\mathcal{H}, \mathcal{K})$), and an operator is said to be M-N-linear if it is both M-linear and N-linear; we denote the collection of all such operators by ${}_M\mathcal{L}_N(\mathcal{H}, \mathcal{K})$.

We shall be concerned primarily with II_1 factors in this book, and it will serve us well to spell out exactly what Theorem 2.1.6 says in this special case.

If M denotes a II_1 factor with separable pre-dual, the symbol tr_M will always denote the unique normal tracial state on M and we shall simply write $L^2(M)$ for $L^2(M, \text{tr}_M)$.

We shall assume throughout this section that M denotes a II_1 factor with separable pre-dual, and we shall write $\mathcal{H}_1 = L^2(M)$ and $\mathcal{H}_\infty = \mathcal{H}_1 \otimes \ell^2$. Consequently, we shall also use the symbols π_1, π_∞ to denote the representations underlying the above M-modules. (Thus, for instance, $\pi_\infty(x) = \pi_1(x) \otimes \text{id}_{\ell^2}$.) Also we shall write $M_\infty(M) = M \otimes \mathcal{L}(\ell^2)$ and think of elements of $M_\infty(M)$ as infinite matrices $((x_{ij}))$ with entries from M .

Notice now that $\mathcal{H}_1$ is actually an M-M-bimodule, with the right action of M being given by $\pi_r(y)\xi = \xi y = Jy^*J\xi$; in fact, Theorem 1.2.4 says that in this case, we have an equality $\pi_1(M)' = \pi_r(M)$.

In order to deal with $\mathcal{H}_\infty$, we shall find it convenient to think of $\mathcal{H}_\infty$ as $\text{Mat}_{1\times\infty}(\mathcal{H}_1)$ – by which we mean the Hilbert space of norm-square-summable sequences with entries from $\mathcal{H}_1$. A moment's thought should convince the reader that $\mathcal{H}_\infty$ is actually an M-$M_\infty(M)$-bimodule with respect to matrix multiplication (explicitly, if $\xi = (\xi_1, \xi_2, \cdots), x \in M, y = ((y_{ij})) \in M_\infty(M)$, and $\eta = x\xi y$, then $\eta_j = \sum_i x\xi_i y_{ij}$), and that in fact, we have $\pi_\infty(M)' = \pi_r(M_\infty(M))$.

Notice now that $M_\infty(M)$ is a II_∞ factor, and the 'faithful normal semifinite trace' Tr on it – as discussed at the end of the last section – is given by the obvious formula $\text{Tr}((p_{ij})) = \sum_{i=1}^\infty \text{tr}_M(p_{ii})$. (This is the 'natural' normalisation to choose, in the sense that $\text{Tr}(1_M \otimes q) = 1$ whenever q is a projection in $\mathcal{L}(\ell^2)$ of rank 1. Throughout the sequel, the symbol Tr – if it is used in the context of $M_\infty(M)$ – will always mean the one defined in this paragraph.)

In the above terminology, the content of Theorem 2.1.6, at least as far as II_1 factors are concerned, may be reformulated thus:

THEOREM 2.2.2 *If $\mathcal{H}$ is any separable M-module, then there exists a projection $p \in M_\infty(M)$ such that $\mathcal{H} \cong \mathcal{H}_\infty p$, and such a projection p is determined uniquely up to Murray–von Neumann equivalence.*

We now come to a fundamental definition.

DEFINITION 2.2.3 *Let $\mathcal{H}$ denote an arbitrary separable module over a II_1 factor M with separable pre-dual. Then define*

$$\dim_M \mathcal{H} = \operatorname{Tr} p$$

where $p \in \mathcal{P}(M_\infty(M))$ is any projection such that $\mathcal{H}_\infty p$ is isomorphic to $\mathcal{H}$ as an M-module.

We shall soon derive some of the basic properties of the assignment $\mathcal{H} \mapsto \dim_M \mathcal{H}$, which should convince the reader that there is every reason to call this quantity the *M-dimension* of the module. Before that, however, we pause to (a) mention an example that might make the reader more favourably disposed to this definition, and (b) make a remark of a historical nature.

EXAMPLE 2.2.4 *Suppose Γ is a discrete subgroup of a semisimple Lie group G such that* $\mathrm{covol}(\Gamma) < \infty$. *Assume further that Γ is an ICC group in the sense discussed in §1.4. Suppose now that π is a 'discrete-series representation' of G, meaning that π is a subrepresentation of the left-regular representation of G, or equivalently, that every (equivalently, that some non-zero) matrix-coefficient of π (i.e., a function on G of the form $\langle \pi(\cdot)\xi, \eta\rangle$ where ξ, η belong to the Hilbert space $\mathcal{H}_\pi$ underlying the unitary representation π) is square-integrable with respect to Haar measure.*

It is then the case that $\pi|_\Gamma$ extends to an isomorphism of $L\Gamma$ onto $\pi(\Gamma)''$; consequently, the algebra $\pi(\Gamma)''$ is also a II_1 factor; and it can be shown – see [GHJ] – that

$$\dim_{\pi(\Gamma)''}(\mathcal{H}_\pi) = \mathrm{covol}(\Gamma) \times d_\pi,$$

where d_π denotes the so-called 'formal dimension' of the discrete series representation π.

REMARK 2.2.5 *We should mention here that what we have termed $\dim_M \mathcal{H}$ occurred first in the work of Murray and von Neumann ([MvN1]) as the so-called 'coupling constant' of the module. Their definition is different from the one presented here (and depends upon a result we do not prove here, since that is not really essential for our purposes). They define this coupling constant as infinity if ${}_M\mathcal{L}(\mathcal{H})$ is an infinite factor, and if this is a factor of type II_1, they show the following is true: pick any $\xi \neq 0$ in $\mathcal{H}$ and let $p \in M$ (resp., $p' \in {}_M\mathcal{L}(\mathcal{H})$) be the projection operator whose range is $[{}_M\mathcal{L}(\mathcal{H})\xi]$ (resp., $[M\xi]$); then the quotient*

$$\frac{\mathrm{tr}_M(p)}{\mathrm{tr}_{({}_M\mathcal{L}(\mathcal{H}))}(p')}$$

turns out to be a positive finite constant which is independent of the initial choice of vector ξ; this is the number that they call the 'coupling constant' of the module, and this number agrees with what we have defined as $\dim_M \mathcal{H}$.

(In order to minimise notation, we assume in the next proposition that $M \subseteq \mathcal{L}(\mathcal{H})$ rather than that $\mathcal{H}$ is some abstract M-module; there is no real distinction here since every unital normal homomorphism of a II_1 factor is an isomorphism onto its image – see the appendix.)

PROPOSITION 2.2.6 *Let $\mathcal{H}$ be a separable Hilbert space, and suppose $M \subseteq \mathcal{L}(\mathcal{H})$ is a II_1 factor.*

(i) For each $d \in [0, \infty]$, there exists an M-module $\mathcal{H}_d$ such that $\dim_M \mathcal{H}_d = d$.

(ii) There exists a unique $d \in [0, \infty]$ such that $\mathcal{H} \cong \mathcal{H}_d$; in particular, two M-modules are isomorphic if and only if they have the same M-dimension.

(iii) $\dim_M \mathcal{H} < \infty \Leftrightarrow M'$ is a II_1 factor.

(iv) $\dim_M L^2(M) = 1$.

(v) If $\{\mathcal{K}_n\}_n$ is any countable collection of separable M-modules, then

$$\dim_M(\bigoplus_n \mathcal{K}_n) = \sum_n \dim_M \mathcal{K}_n.$$

(vi) If $\dim_M \mathcal{H} < \infty$ so that M' is a II_1 factor (see (iii) above), and if $p' \in \mathcal{P}(M')$, then $\dim_M(p'\mathcal{H}) = \mathrm{tr}_{M'}(p') \dim_M \mathcal{H}$.

(vii) $p \in \mathcal{P}(M) \Rightarrow \dim_{M_p}(p\mathcal{H}) = (\mathrm{tr}_M(p))^{-1} \dim_M \mathcal{H}$.

(viii) If $\dim_M \mathcal{H} < \infty$ – see (iii) above – then

$$\dim_{M'} \mathcal{H} = (\dim_M \mathcal{H})^{-1}.$$

Proof: The first two assertions follow immediately from Theorem 2.2.2, and from the facts concerning the semifinite trace Tr that were listed out in §1.1.

We shall find the following description of $\mathcal{H}_d, d < \infty$, convenient.

For $d = n \in \mathbb{N}, \mathcal{H}_n$ is the direct sum of n copies of $\mathcal{H}_1$. As in the case of $d = \infty$, we think of $\mathcal{H}_n$ as $\mathrm{Mat}_{1\times n}(\mathcal{H}_1)$; observe, as before, that $\mathcal{H}_n$ is naturally an M-$M_n(M)$-bimodule (with respect to matrix multiplication) in such a way that ${}_M\mathcal{L}(\mathcal{H}_n) = \pi_r(M_n(M))$.

If $d \in [0, \infty)$, pick an integer n which is at least as large as d, pick a projection q in the II_1 factor $M_n(M)$ such that $\mathrm{tr}_{M_n(M)}(q) = \frac{d}{n}$ and set $\mathcal{H}_d = \mathcal{H}_n q$.

(iii) Notice that if $\mathcal{H} = \mathcal{H}_\infty p$, for some $p \in \mathcal{P}(M_\infty(M))$, then $M' = M_\infty(M)_p$, and that $\mathrm{Tr}\, p < \infty \Leftrightarrow p$ is a finite projection.

(iv) If $p = 1_M \otimes e_{11}$, where e_{11} denotes the matrix-unit (with 1 in the (1,1)-place and zeros elsewhere), then $\mathcal{H}_\infty p \cong L^2(M)$.

(v) Let $\dim_M \mathcal{K}_n = d_n$. We can find mutually orthogonal projections p_n in $M_\infty(M)$ such that $\mathrm{Tr}\, p_n = d_n$ for each n. If $p = \sum_n p_n$, the desired assertion follows from the fact that Tr is 'countably additive' on $\mathcal{P}(M_\infty(M))$, since $\bigoplus_n \mathcal{K}_n \cong \mathcal{H}_\infty p$.

(vi) We may as well assume that $\mathcal{H} = \mathcal{H}_d$, where $\mathcal{H}_d$ has been constructed as in the discussion preceding the proof of (iii) of this proposition. Let n, q have the same meanings as above. In this case, since ${}_M\mathcal{L}(\mathcal{H}_d) = M_n(M)_q$, it follows that there exists a projection $p \in M_n(M)$ such that $p \leq q$ and $p' = \pi_r(p)$. Then, $p'\mathcal{H} = \mathcal{H}_n p$ and consequently, we see that

$$\dim_M(p'\mathcal{H}) = n\,\mathrm{tr}_{M_n(M)}(p).$$

On the other hand, it follows from the uniqueness of the trace in a II_1 factor that

$$\begin{aligned}\mathrm{tr}_{M'}(p') &= \mathrm{tr}_{M_n(M)_q}(p)\\ &= \frac{\mathrm{tr}_{M_n(M)}(p)}{\mathrm{tr}_{M_n(M)}(q)}\\ &= \frac{\frac{1}{n}\dim_M(p'\mathcal{H})}{\frac{1}{n}\dim_M\mathcal{H}}.\end{aligned}$$

(vii) Notice that the conclusion is 'additive in $\mathcal{H}$', in the sense that if $\mathcal{H} = \bigoplus_n \mathcal{H}_n$ is a decomposition of $\mathcal{H}$ into countably many M-submodules, and if the desired assertion is valid for each $\mathcal{H}_n$, then the desired assertion is valid for $\mathcal{H}$ as well. Hence there is no loss of generality in assuming that $\dim_M \mathcal{H} \leq \mathrm{tr}_M p (\leq 1)$.

Hence – see the discussion just prior to the proof of (iii) above – we may assume that $\mathcal{H} = (\mathcal{H}_1)q$, where $q \in \mathcal{P}(M)$ and $q \leq p$. It is then seen that

$$\begin{aligned}\dim_{M_p}(p\mathcal{H}) &= \dim_{M_p}(p(\mathcal{H}_1)q)\\ &= \dim_{M_p}(p(\mathcal{H}_1)p \cdot q)\\ &= \dim_{M_p}(L^2(M_p)q)\\ &= \mathrm{tr}_{M_p}(q)\cdot 1 \qquad \text{(by (vi) and (iv))}\\ &= \frac{\mathrm{tr}_M q}{\mathrm{tr}_M p}\\ &= \frac{\dim_M \mathcal{H}}{\mathrm{tr}_M p}.\end{aligned}$$

(viii) We may assume that $\mathcal{H} = \mathcal{H}_d$ is constructed as described in the comments preceding the proof of (iii) above. First, if $d = 1$, this is a consequence of Theorem 1.2.4 (1). Next, if $d = n$, note that $\mathcal{H}_n e_{11} = \mathcal{H}_1$ and that $\pi_r(e_{11})$ is a projection in $\pi_r(M_n(M)) = \pi_l(M)'$ with trace $\frac{1}{n}$; it follows from (vii) above that

$$\begin{aligned}1 &= \dim_{({}_M\mathcal{L}(\mathcal{H}_1))}\mathcal{H}_1\\ &= \dim_{({}_M\mathcal{L}(\mathcal{H}_n)_{\pi_r(e_{11})})}(\mathcal{H}_n e_{11})\\ &= \frac{1}{\frac{1}{n}}\dim_{({}_M\mathcal{L}(\mathcal{H}_n))}(\mathcal{H}_n);\end{aligned}$$

hence $\dim_{({}_M\mathcal{L}(\mathcal{H}_n))}(\mathcal{H}_n) = \frac{1}{n}$, thus proving the assertion when $d = n$. A similar reasoning, applied to the fact that $\mathcal{H}_d$ is obtained by 'cutting down' $\mathcal{H}_n$, proves the general case. □

2.3 Subfactors and index

We begin with a look at some examples of bimodules (or correspondences, as introduced and popularised by Connes – see [Con2]).

Suppose, to be specific, that M is a II_1 factor and that $\mathcal{H}_1 = L^2(M)$. Then, as has been already observed, $\mathcal{H}_1$ is an M-M-bimodule in a natural fashion; but there are several other bimodule structures on $L^2(M)$ as follows. Fix an automorphism θ of M, and define the bimodule $\mathcal{H}_\theta$ to be $L^2(M)$ as a Hilbert space, but where the actions are given by: $x \cdot \xi \cdot y = x\xi\theta(y)$, where the actions in the right side are as in $\mathcal{H}_1$ (and the actions on the left refer to the actions in $\mathcal{H}_\theta$). It is easy to see that $\mathcal{H}_\theta$, so defined, is indeed an M-M-bimodule, which is *irreducible* in the sense that $\mathcal{H}_\theta$ has no proper sub-bimodules (or equivalently, the von Neumann algebra ${}_M\mathcal{L}_M(\mathcal{H}_\theta)$ reduces to the scalar multiples of the identity operator). Thus, there are plenty of irreducible M-M-bimodules, although there are no irreducible M-modules.

Further, since the only left-M-linear maps on $L^2(M)$ are of the form $\pi_r(x), x \in M$, it follows that if $\theta_i \in \operatorname{Aut} M, i = 1, 2$, then the bimodules $\mathcal{H}_{\theta_i}$ are isomorphic if and only if the automorphisms are *outer equivalent* – meaning that there exists a unitary element $u \in M$ such that $\theta_1(x) = \theta_2(uxu^*)\ \forall x \in M$. Thus, the set of equivalence classes of irreducible M-M-bimodules is at least as rich as the group of outer automorphisms (viz., the quotient of $\operatorname{Aut} M$ by the normal subgroup of inner automorphisms). In fact, it is much richer, as can be seen, for instance, from the fact that we could have defined the bimodule $\mathcal{H}_\theta$ by only requiring θ to be a unital *endomorphism*. (It is a fact that if M is a factor of type III, then every M-M-bimodule is isomorphic to $\mathcal{H}_\theta$, for some endomorphism θ – see the third paragraph of §4.1.)

It is clear that the assignment

$$\mathcal{H} \mapsto (\dim_{M-}(\mathcal{H}), \dim_{-M}(\mathcal{H})) \tag{2.3.1}$$

defines an isomorphism-invariant of the M-M-bimodule $\mathcal{H}$. The bimodules $\mathcal{H}_\theta$ show that the above is not a complete invariant of the bimodule. Motivated by the success of Theorem 2.1.6, we are naturally led to the following question.

QUESTION 2.3.1 *What can be said about the set*

$$\{(\dim_{M-}(\mathcal{H}), \dim_{-M}(\mathcal{H})) : \mathcal{H} \text{ is a separable } M\text{-}M\text{-}bimodul\}?$$

Before we get to this question, pause to notice that if M is a factor and if $\mathcal{H}$ is an M-M-bimodule, then, by definition of a bimodule, we see that $\pi_l(M)$ is a subfactor of $\pi_r(M)'$; further, we see that, at least in this example, the subfactor has 'trivial relative commutant' in the ambient factor precisely when the initial bimodule is irreducible. We formalise all of this in the next definition.

DEFINITION 2.3.2 *A subfactor of a factor M is a subalgebra $N \subseteq M$ such that N is also a factor and N contains the identity element of M. The subfactor N is said to be irreducible if it has 'trivial relative commutant' – i.e., if $N' \cap M = \mathbb{C}1$.*

EXAMPLE 2.3.3 *(a) Suppose Γ_0 is a subgroup of Γ, and suppose that both Γ and Γ_0 are ICC groups. Notice then that $L\Gamma_0$ sits naturally as a-subfactor of the II_1 factor $L\Gamma$. Notice, further, that $\ell^2(\Gamma) = L^2(L\Gamma)$ and that if $\Gamma = \coprod \Gamma_0 s_i$ is the partition of Γ into disjoint cosets of Γ_0, then $\ell^2(\Gamma) = \bigoplus(\ell^2(\Gamma_0))s_i$ is an orthogonal decomposition of $\ell^2(\Gamma)$ into (left) $L\Gamma_0$- submodules (each of which is isomorphic to $\ell^2(\Gamma_0)$ as an $L\Gamma_0$-module, and consequently, we have*

$$\dim_{L\Gamma_0}(\ell^2(\Gamma)) = [\Gamma : \Gamma_0].$$

(b) Suppose G is a finite group acting on a II_1-factor P. It is true – see the first paragraph of §A.4 – that an automorphism θ of a II_1 factor is 'free' in the sense of Definition 1.4.2 if and only if it is not an inner automorphism (i.e., not of the form $\mathrm{Ad}\, u = u(\cdot)u^*$ for some unitary u in the algebra). It follows that if P, G are as above, then the crossed product algebra $P \times G$ is a II_1 factor if and only if the action is outer *– meaning that no non-identity element of G acts as an inner automorphism.*

Suppose then that G acts as outer automorphisms of P, so that $M = P \times G$ is a II_1 factor. It should be clear now that every subgroup H of G would yield a subfactor $N = P \times H$ of M. Notice now that there is a natural identification $L^2(M) = L^2(P) \otimes \ell^2(G)$ and that if $G = \coprod H s_i$ is the partition of G into distinct cosets of H, then (i) each of the subspaces $\mathcal{H}_i = L^2(P) \otimes [\{\xi_{hs_i} : h \in H\}]$ is an N-submodule of $L^2(M)$, which is isomorphic, as an N-module, to $L^2(N)$, and (ii) $L^2(M) = \bigoplus \mathcal{H}_i$. It follows that $\dim_N(L^2(M)) = [G : H]$.

Motivated by the preceding examples, we make the following definition.

DEFINITION 2.3.4 *If N is a subfactor of a II_1 factor M, define the index of N in M by the expression*

$$[M : N] = \dim_N(L^2(M)).$$

The next proposition shows that the index $[M : N]$ can be read off from any M-module of finite M-dimension.

PROPOSITION 2.3.5 *Let $N \subseteq M$ be an inclusion of II_1 factors. Let $\mathcal{H}$ be any separable M-module such that $\dim_M \mathcal{H} < \infty$. Then*

$$\dim_N \mathcal{H} < \infty \Leftrightarrow [M : N] < \infty;$$

in fact, we have the identity

$$\dim_N \mathcal{H} = [M : N] \dim_M \mathcal{H}. \tag{2.3.2}$$

Proof: Suppose, to start with, that $\mathcal{K}_i, i = 1, 2$, are any two M-modules with $\dim_M \mathcal{K}_i < \infty$. It follows then that there exists $n \in \mathbb{N}$ and a projection $q' \in {}_M\mathcal{L}(\mathcal{K}_2 \otimes \mathbb{C}^n)$ such that $\mathcal{K}_1 \cong q'(\mathcal{K}_2 \otimes \mathbb{C}^n)$; thus we may deduce that $\dim_N \mathcal{K}_1 \leq \dim_N(\mathcal{K}_2 \otimes \mathbb{C}^n) = n \times \dim_N \mathcal{K}_2$. Since the roles of $\mathcal{K}_1$ and $\mathcal{K}_2$ are interchangeable, we see that $\dim_N \mathcal{K}_1 < \infty \Leftrightarrow \dim_N \mathcal{K}_2 < \infty$. In particular, we see that $\dim_N \mathcal{H} < \infty \Leftrightarrow [M : N] < \infty$.

To prove the asserted equality, we may assume that, as an M-module, $\mathcal{H} = \mathcal{H}_n q$, where $\mathcal{H}_n = L^2(M) \oplus \overset{n \text{ copies}}{\cdots} \oplus L^2(M), q \in M_n(M), \mathrm{tr}_{M_n(M)} q = \frac{1}{n} \dim_M \mathcal{H}$. Then $\dim_N \mathcal{H}_n < \infty$ so $N'(= {}_N\mathcal{L}(\mathcal{H}_n))$ is a II_1 factor, and hence, it follows that

$$\begin{aligned} \dim_N \mathcal{H} &= \mathrm{tr}_{N'} \pi_r(q) \dim_N \mathcal{H}_n \\ &= \mathrm{tr}_{M'} \pi_r(q) n \dim_N \mathcal{H}_1 \\ &= \dim_M \mathcal{H}[M : N]. \end{aligned}$$ □

Before proceeding further, we record an immediate consequence of this fact and Proposition 2.2.6(viii).

COROLLARY 2.3.6 *(a) If $N \subseteq M$ is an inclusion of II_1 factors, and if $M \subseteq \mathcal{L}(\mathcal{H})$ and $\dim_M \mathcal{H} < \infty$, then*

$$[N' : M'] = [M : N].$$

(b) If $N \subseteq M \subseteq P$ is a tower of II_1 factors, then

$$[P : N] = [P : M][M : N].$$ □

In particular, if $\mathcal{H}$ is an M-M-bimodule which is *bifinite* in the sense that both $\dim_{M-} \mathcal{H} < \infty$ and $\dim_{-M} \mathcal{H} < \infty$, it follows then that

$$\dim_{M-} \mathcal{H} \cdot \dim_{-M} \mathcal{H} = [\pi_r(M)' : \pi_l(M)].$$

Thus, we find that the answer to Question 2.3.1 is tied up with the following related question.

QUESTION 2.3.7 *What are the possible values of $[M : N]$, where $N \subseteq M$ is an inclusion of II_1 factors ?*

The following result – see [Jon1] – completely answers Question 2.3.7, and the answer is quite unexpected in the light of our experience with the classification of modules.

THEOREM 2.3.8 *If $N \subseteq M$ is any inclusion of II_1 factors, then*

$$[M : N] \in \{4 \cos^2 \frac{\pi}{n} : n = 3, 4, \cdots\} \cup [4, \infty].$$

Further, if $\lambda \in (\{4 \cos^2 \frac{\pi}{n} : n = 3, 4, \cdots\} \cup [4, \infty])$, then there exists a subfactor R_λ of the hyperfinite II_1 factor such that $[R : R_\lambda] = \lambda$.

In this section, we shall content ourselves with describing a construction of subfactors of all possible index values ≥ 4.

LEMMA 2.3.9 *Let $N \subseteq M$ be an inclusion of II_1 factors, and suppose $M \subseteq \mathcal{L}(\mathcal{H})$ where $\dim_M \mathcal{H} < \infty$. If $p \in \mathcal{P}(N' \cap M)$, then*

$$[pMp : Np] = \mathrm{tr}_{N'}(p) \cdot \mathrm{tr}_M(p) \cdot [M : N].$$

Proof: It follows from Proposition 2.3.5 that

$$\begin{aligned} [pMp : Np] &= \frac{\dim_{Np}(p\mathcal{H})}{\dim_{pMp}(p\mathcal{H})} \\ &= (\mathrm{tr}_{N'}(p) \dim_N \mathcal{H}) \cdot \dim_{M'p}(p\mathcal{H}) \\ &= (\mathrm{tr}_{N'}(p) \dim_N \mathcal{H}) \cdot (\mathrm{tr}_M(p) \dim_{M'} \mathcal{H}), \end{aligned}$$

as desired. □

PROPOSITION 2.3.10 *Suppose M is a II_1 factor such that $M_p \cong M_{1-p}$ for some projection $p \in M$. Let $\theta : M_p \to M_{1-p}$ be such an isomorphism, and define $N = \{x + \theta(x) : x \in M_p\}$. Then N is a subfactor of M with $[M : N] = \frac{1}{\mathrm{tr}_M p} + \frac{1}{\mathrm{tr}_M (1-p)}$.*

Proof: Note that $p \in N' \cap M$ and that $pMp = Np$, so that $[pMp : Np] = 1$. Hence, by Lemma 2.3.9,

$$1 = \mathrm{tr}_{N'}(p) \cdot \mathrm{tr}_M(p) \cdot [M : N],$$

and so

$$\frac{1}{\mathrm{tr}_M(p)} = \mathrm{tr}_{N'}(p) \cdot [M : N].$$

Similarly, we also have

$$\frac{1}{\mathrm{tr}_M(1-p)} = \mathrm{tr}_{N'}(1-p) \cdot [M : N].$$

Add the above equations to obtain the desired conclusion. □

The existence of subfactors of the hyperfinite II_1 factor R with index at least 4 is an immediate consequence of proposition 1.4.9 and the preceding proposition (and the trivial fact that the mapping $t \mapsto (\frac{1}{t} + \frac{1}{1-t})$ maps (0,1) onto $(0, \infty)$).

The preceding construction used the existence of non-trivial projections in the relative commutant. The following question, which naturally arises, still remains unanswered.

QUESTION 2.3.11 *Describe the set of index values $[R : R_0]$ of irreducible subfactors R_0 of the hyperfinite factor R.*

We conclude this section with some useful consequences of Lemma 2.3.9.

PROPOSITION 2.3.12 *Suppose $N \subseteq M$ is an inclusion of II_1 factors such that $[M : N] < \infty$. If there exist pairwise orthogonal non-zero projections $p_1, \cdots, p_n \in N' \cap M$, then $[M : N] \geq n^2$.*

In particular,

(i)$[M : N] < \infty \Rightarrow N' \cap M$ is finite-dimensional;

(ii)$[M : N] < 4 \Rightarrow N' \cap M = \mathbb{C}1$.

Proof: Deduce from Lemma 2.3.9 that

$$\begin{aligned} [M : N] &\geq \sum_{i=1}^{n} \operatorname{tr}_M(p_i)[M : N] \\ &\geq \sum_{i=1}^{n} \frac{1}{\operatorname{tr}_{N'}(p_i)} \\ &\geq n^2 \end{aligned}$$

(since $r_1, \cdots, r_m \in (0,1), \sum_i r_i = 1 \Rightarrow \sum_{i=1}^{m} \frac{1}{r_i} \geq m^2$).

The above inequality clearly implies (i) and (ii). □

Chapter 3

Some basic facts

3.1 The basic construction

Suppose $N \subseteq M$ is an inclusion of finite von Neumann algebras. Fix some faithful normal tracial state tr on M, and consider the M-M-bimodule $\mathcal{H} = L^2(M, \text{tr})$, with its distinguished cyclic trace vector Ω. Since $\mathcal{H}$ is the completion of $M\Omega$, it follows that the subspace $\mathcal{H}_1 = [N\Omega]$ of $\mathcal{H}$ can be naturally identified with $L^2(N, \text{tr})$. Let e_N denote the orthogonal projection of $\mathcal{H}$ onto the subspace $\mathcal{H}_1$. It is a consequence of Theorem 1.2.4(1) that $e_N(M\Omega) \subseteq N\Omega$, and hence the projection e_N induces, by restriction, a map $E : M \to N$. The map E is called the (tr-preserving) **conditional expectation** of M onto N, and is easily seen to satisfy the following properties:

(i)
$$e_N x e_N = E(x) e_N \ \forall x \in M, \tag{3.1.1}$$
and consequently E defines a (Banach space) projection of M onto N.

(ii) E is N-N-bilinear, meaning that $E(n_1 m n_2) = n_1 E(m) n_2$.

(iii) $\text{tr} \circ E = \text{tr}$.

It is clear from equation (3.1.1) that the conditional expectation is a *-preserving map, or, in other words, that

$$J e_N = e_N J, \tag{3.1.2}$$

where J denotes the modular conjugation operator on $\mathcal{H}$.

DEFINITION 3.1.1 *The passage from the initial inclusion $N \subseteq M$ to the von Neumann algebra $\langle M, e_N \rangle = (M \cup \{e_n\})''$, and consequently to the tower $N \subseteq M \subseteq \langle M, e_N \rangle$, is called the basic construction.*

We list some simple relations between the various objects involved in the basic construction. Also, we adopt the convention that M (and hence, also N) is identified, via π_l, with a subalgebra of $\mathcal{L}(\mathcal{H})$; thus, for instance, the right action of M on $\mathcal{H}$ is given by $\pi_r(x) = J x^* J$.

PROPOSITION 3.1.2 *With the foregoing notation, we have:*

(i) $e_N \in N'$.

(ii) $N = M \cap \{e_N\}'$.

(iii) $\langle M, e_N \rangle = JN'J$.

(iv) Assume both M *and* N *are* II_1 *factors. Then:*

(a) $\langle M, e_N \rangle$ *is a* II_1 *factor if and only if* $[M : N] < \infty$*; in this case, we have* $[\langle M, e_N \rangle : M] = [M : N]$.

(b) $\mathrm{tr}_{\langle M, e_N \rangle}(e_N) = [M : N]^{-1}$.

(c) (Markov property) $E_M(e_N) = \mathrm{tr}_{\langle M, e_N \rangle}(e_N)$ *– where* E_M *denotes the conditional expectation of the* II_1 *factor* $\langle M, e_N \rangle$ *onto the subfactor* M.

Proof: The first assertion is obvious; as for the second, the fact that the trace vector Ω is separating for M implies that the map $x \mapsto xe_N$ is injective, and consequently, if $x \in M$, it follows from equation (3.1.1) that x commutes with e_N if and only if $x = Ex$.

(iii) It follows from (ii) that

$$\begin{aligned} JN'J &= J\langle M', \{e_N\}'' \rangle J \\ &= \langle JM'J, Je_NJ \rangle \\ &= \langle M, e_N \rangle. \end{aligned}$$

(iv) (a) The first assertion is a consequence of (iii) above and Proposition 2.2.6(iii). As for the second, note that

$$\begin{aligned} [\langle M, e_N \rangle : M] &= (\dim_{\langle M, e_N \rangle} L^2(M))^{-1} \\ &= (\dim_{JN'J} \mathcal{H})^{-1} \\ &= (\dim_{N'} \mathcal{H})^{-1} \\ &= \dim_N \mathcal{H} \\ &= [M : N]. \end{aligned}$$

(b) On the one hand, $\mathrm{tr}_{\langle M, e_N \rangle} e_N = \mathrm{tr}_{N'} e_N$, while on the other,

$$\begin{aligned} 1 &= \dim_N L^2(N) \\ &= \dim_N(e_N \mathcal{H}) \\ &= \mathrm{tr}_{N'} e_N \cdot [M : N]. \end{aligned}$$

(c) We need to verify that

$$\mathrm{tr}_{\langle M, e_N \rangle}(xe_N) = \tau \mathrm{tr}_M(x) \ \forall x \in M,$$

where $\tau = \mathrm{tr}_{\langle M, e_N \rangle}(e_N) = [M : N]^{-1}$. We first verify this identity whenever $x \in N$; for this, note that, in view of (i), the map $x \mapsto \mathrm{tr}(xe_N)$ is a trace on the II_1 factor N which takes the value τ at the identity; appeal to the uniqueness of the trace in a finite factor to conclude the proof in the special case. The case of a general $x \in M$ reduces to the above case because of equation (3.1.1), thus:

$$\begin{aligned} \mathrm{tr}_{\langle M,e_N\rangle}(xe_N) &= \mathrm{tr}_{\langle M,e_N\rangle}(xe_N^2) \\ &= \mathrm{tr}_{\langle M,e_N\rangle}(e_N x e_N) \\ &= \mathrm{tr}_{\langle M,e_N\rangle}(E_N(x)e_N). \end{aligned}$$ □

REMARK 3.1.3 *In the sequel, if $N \subseteq M$ is an inclusion of finite von Neumann algebras, if* tr *is a fixed tracial state on M, if $\langle M, e_N\rangle$ is as above, and if ϕ denotes an extension of* tr *to $\langle M, e_N\rangle$, we shall say that e_N satisfies the Markov property with respect to the algebra M (and ϕ) if it is the case that $E_M(e_N) = \phi(e_N)$, where E_M denotes the unique ϕ-preserving conditional expectation of $\langle M, e_N\rangle$ onto M.*

Before proceeding further, we record two simple consequences of the last proposition.

COROLLARY 3.1.4 *Suppose $N \subseteq M$ is an inclusion of II_1 factors. Then*

$$[M : N] = 1 \Leftrightarrow N = M.$$

Proof: If e_N and $\langle M, e_N\rangle$ are as above, then the assumption $[M : N] = 1$ implies that $\mathrm{tr}_{\langle M,e_N\rangle} e_N = 1$, and the faithfulness of the trace now implies that $e_N = 1$; thus, in the notation of the first paragraph of this section, we have $\mathcal{H}_1 = \mathcal{H}$. An appeal to Proposition 3.1.2 (i) completes the proof. □

COROLLARY 3.1.5 *If $N \subseteq M$ is an inclusion of II_1 factors, then*

$$[M : N] \notin (1, 2).$$

Proof: On the one hand, it follows from (iv)(a),(b) of Proposition 3.1.2 that $[\langle M, e_N\rangle : N] = [M : N]^2$. On the other hand, since $e_N \in N' \cap \langle M, e_N\rangle$, the non-triviality of the relative commutant implies, via Proposition 2.3.12(ii), that we must have $[M : N] \geq 2$. □

We conclude this section with one way of constructing subfactors of the hyperfinite II_1 factor.

EXAMPLE 3.1.6 *Let us take the model $R = R_{(2)}$ (in the notation of §1.4) of the hyperfinite II_1 factor. (Everything we say can be said just as well with any N in place of 2.) Let A_n have the same meaning as in the above-mentioned construction of $R_{(N)}$. Notice, to start with, that, for any n, there is an obvious identification $A_n \otimes A_2 \cong A_{n+2}$; for any $x \in A_2$, let us denote the image of*

$1 \otimes x$ under this identification by $x_{n+1,n+2}$. (Thus, $x_{n+1,n+2}$ is just the element x 'sitting in slots $n+1, n+2$'.)

Now fix a unitary element $u \in A_2$, and notice that the equation

$$\theta_u(x) = \lim_{n \to \infty} \mathrm{Ad}_{u_{12}u_{23}\cdots u_{n,n+1}}(x) \tag{3.1.3}$$

defines a unital normal endomorphism θ_u of $R = R_{(2)}$. Consider the subfactor $R_u = \theta_u(R)$ of R. Thus, we get a family of subfactors of R parametrised by the unitary group $U(4, \mathbb{C})$. It must be clear that the subfactor R_1, corresponding to $u = 1$, is the trivial subfactor $R_1 = R$. On the other hand, if we take u to be the unitary operator implementing the 'flip' on $\mathbb{C}^2 \otimes \mathbb{C}^2$ – or, in other words,

$$u = \begin{bmatrix} 1 & 0 & 0 & 0 \\ 0 & 0 & 1 & 0 \\ 0 & 1 & 0 & 0 \\ 0 & 0 & 0 & 1 \end{bmatrix}$$

– then it is easy to see that θ_u defines the 'shift', thus:

$$\theta_u(x_1 \otimes x_2 \otimes \cdots) = 1 \otimes x_1 \otimes x_2 \otimes \cdots$$

whenever $x_1, x_2, \cdots \in M_2(\mathbb{C})$. Thus, we find that in this case, $R = M_2(\mathbb{C}) \otimes R_u$, and consequently, $[R : R_u] = 4$.

Since the unitary group $U(4, \mathbb{C})$ is connected, we find that 'the index is a discontinuous function of the subfactor'!

3.2 Finite-dimensional inclusions

In this section, we recall various elementary facts concerning finite-dimensional von Neumann algebras.

To start with, any finite-dimensional C^*-algebra A is semi-simple and hence, by the Wedderburn–Artin theorem, is isomorphic to the direct sum of finitely many matrix algebras over $\mathbb{C}$. To be specific, if A is a finite-dimensional C^*-algebra, then

$$A \cong M_{n_1}(\mathbb{C}) \oplus M_{n_2}(\mathbb{C}) \oplus \cdots \oplus M_{n_k}(\mathbb{C}) \tag{3.2.1}$$

for a uniquely determined integer $k(= \dim Z(A))$ and a uniquely determined set $\{n_1, \cdots, n_k\}$ of positive integers. (The fact is that A admits exactly k equivalence classes of (pairwise inequivalent) irreducible representations, and the dimensions of these representations are precisely the integers $n_1, \cdots, n_k$.)

If equation (3.2.1) is satisfied, we shall say that A is *of type* $(n_1, \cdots, n_k)$ and we shall refer to the vector $\overline{n} = (n_1, \cdots, n_k)$ as the *dimension vector* of A.

Further, since a matrix algebra is a factor, it follows that if A is as above, then there is a bijective correspondence between the set of faithful tracial states on A and the open simplex $\Delta^k = \{\bar{t} = (t_1, \cdots, t_k)' \in I\!R^k : t_j > 0 \ \forall j$ and $\bar{n}\bar{t} = 1\}$ given by $\tau \mapsto \bar{t}$, where $t_j = \tau(p_j)$ is the trace of a minimal (= rank one) projection in the j-th summand of A. (The reason for the superscript $'$ (which denotes 'transpose') is that we shall find it convenient to adopt the convention that dimension-vectors are row-vectors while trace-vectors are column-vectors.)

That's about all there is to say about finite-dimensional C^*- (and hence von Neumann) algebras.

Suppose now that we are given an inclusion $A \subseteq B$ of finite-dimensional C^*-algebras. Then, besides the dimension-vectors, say $\bar{n} = (n_1, \cdots, n_k)$ and $\bar{m} = (m_1, \cdots m_l)$, of A and B, respectively, there is more data needed to describe how the smaller algebra is included in the larger. The more that is needed is the $k \times l$ **inclusion matrix** $\Lambda = \Lambda_A^B$ described thus: Λ_{ij} is the number of times that the i-th irreducible representation of A features, in the restriction, to A, of the j-th irreducible representation of B. Less formally, but more transparently, this is the number of times the i-th summand of A gets repeated in the j-th summand of B. This data can clearly be encoded equally efficiently in a bipartite (multi-) graph with k even and l odd vertices, where the i-th even vertex is joined to the j-th odd vertex by Λ_{ij} bonds. (This graph is sometimes referred to as the **Bratteli diagram** of the inclusion $A \subseteq B$.) It is a fact that the isomorphism-class of the inclusion is uniquely determined by the inclusion matrix and the dimension vectors of A and B.

EXAMPLE 3.2.1 *The inclusion*

$$\begin{array}{rcl} A & = & \left\{\left(\begin{bmatrix} X & 0 \\ 0 & y \end{bmatrix}, \begin{bmatrix} y & 0 \\ 0 & y \end{bmatrix}, [y]\right) : X \in M_2(\mathbb{C}), y \in \mathbb{C}\right\} \\ \cap & & \\ B & = & M_3(\mathbb{C}) \oplus M_2(\mathbb{C}) \oplus M_1(\mathbb{C}) \end{array}$$

is described by the data

$$\begin{array}{cccc} A & \bar{n} & = & (2,1) \\ \cap & \Lambda_A^B & = & \begin{bmatrix} 1 & 0 & 0 \\ 1 & 2 & 1 \end{bmatrix} \\ B & \bar{m} & = & (3,2,1) \end{array}$$

Bratteli diagram

□

It must be clear from the definitions that if the inclusion is unital – i.e., if the subalgebra contains the identity of the big algebra, and this is the only kind of inclusion that we shall ever consider here – then the dimension-vectors are related to the inclusion matrix by the equation

$$\bar{m} = \bar{n}\Lambda. \tag{3.2.2}$$

Similarly, if a tracial state τ on A (resp., σ on B) corresponds to the 'trace-vector' $\overline{t}$ (resp., $\overline{s}$), then

$$\tau = \sigma|_A \Leftrightarrow \overline{t} = \Lambda \overline{s}. \tag{3.2.3}$$

It should be clear that if we had a whole tower

$$A_1 \subseteq A_2 \subseteq \cdots \subseteq A_n \subseteq A_{n+1} \subseteq \cdots$$

of finite-dimensional C^*-algebras, then we could stack up the Bratteli diagrams corresponding to the several inclusions and obtain the Bratteli diagram of the tower. (Of course, for this to be possible, one must consistently label the direct summands of any algebra in the tower.) Or, in terms of the inclusion matrices, one should observe that – with the summands of the A_k's consistently labelled – $\Lambda_{A_n}^{A_{n+2}} = \Lambda_{A_n}^{A_{n+1}} \cdot \Lambda_{A_{n+1}}^{A_{n+2}}$.

For instance, we might have the following tower:

$$A_1 \subseteq A_2 \subseteq A_3 \subseteq A_4 \subseteq A_5 \subseteq \cdots$$

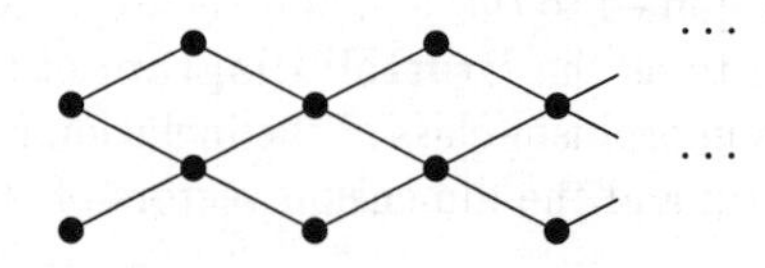

A good feature about a tower $\{A_n\}_{n=1}^\infty$, as above, is that it has 'finite width' – meaning that $\sup_n \dim Z(A_n) < \infty$ – and the Bratteli diagram is connected and is periodic (of order two). The reason this is 'good' is that for such a tower, the algebra $\bigcup A_n$ admits a unique tracial state and yields the hyperfinite II_1 factor as its completion in the GNS representation associated with the trace – as in our discussion of $R_{(N)}$ in §1.4. (This is a consequence of the general fact that if $\{A_n\}_{n=1}^\infty$ is an arbitrary tower of C^*-algebras and if τ is a tracial state on $A_\infty = \bigcup A_n$, and if π_τ denotes the GNS representation associated with τ, then $\pi_\tau(A_\infty)''$ is a factor if and only if τ is an extreme point in the set of tracial states of A_∞.)

We now consider the basic construction, when applied to an inclusion $A \subseteq B$ of finite-dimensional C^*-algebras. Notice that $\langle B, e_A \rangle$ is, by definition, an algebra of operators on the finite-dimensional space $L^2(B, \sigma)$, and is hence also finite-dimensional.

LEMMA 3.2.2 *(a) Let $A \subseteq B$ be a unital inclusion of finite-dimensional C^*-algebras. Fix a faithful tracial state ϕ on B, and let J denote the modular conjugation operator on $L^2(B, \phi)$. Let e denote the orthogonal projection of $L^2(B, \phi)$ onto the subspace $L^2(A, \phi|_A)$; let E_A denote the ϕ-preserving conditional expectation of B onto A; and let $B_1 = \langle B, e \rangle$ denote the result of the* *basic construction. Then:*

(i) $ze = Jz^*Je, \ \forall z \in Z(A)$;

(ii) *if* $b \in B$, *then* $b \in Z(B) \Leftrightarrow b = Jb^*J$;

(iii) $z \mapsto Jz^*J$ *is a *-isomorphism of* $Z(A)$ *onto* $Z(B_1)$;

(iv) *if* $p_1, \cdots, p_m$ *(resp.,* $q_1, \cdots, q_n$*) is an ordering of the set of minimal central projections of* A *(resp.,* B*), and if we set* $\tilde{p}_i = Jp_iJ, 1 \leq i \leq m$, *then* $\tilde{p}_1, \cdots, \tilde{p}_m$ *is an ordering of the minimal central projections of* B_1; *if* Λ *is the inclusion matrix for* $A \subseteq B$, *computed with respect to the* p_i*'s and* q_j*'s, then the inclusion matrix for* $B \subseteq B_1$, *computed with respect to the* q_j*'s and the* $\tilde{p}_i$*'s, is the transpose matrix* Λ'.

(b) *Furthermore,*

(i) *the map* $a \mapsto ae$ *is a *-isomorphism of* A *onto* eB_1e;

(ii) *if* p *is a minimal central projection in* A, *and if* p_0 *is a minimal projection of* A *such that* $p_0 \leq p$, *then* p_0e *is a minimal projection in* B_1 *which is majorised by the minimal central projection* JpJ *of* B_1.

Proof: (a)(i) For all $z \in Z(A), b \in B$, we have

$$zeb\Omega_B = zE_A(b)\Omega_B = E_A(b)z\Omega_B = Jz^*JE_A(b)\Omega_B = Jz^*Jeb\Omega_B.$$

(ii) This is obvious.

(iii) The map $a \to Ja^*J$ is an anti-isomorphism of A into $\mathcal{L}(L^2(B))$, and hence its restriction to the abelian subalgebra $Z(A)$ is an isomorphism onto $J(A \cap A')J = B_1' \cap B_1$.

(iv) Only the assertion about the inclusion matrices needs proof. If $\tilde{\Lambda}$ denotes the inclusion matrix for $B \subseteq B_1$, the definitions imply that

$$\Lambda_{ij} = [\dim_{\mathbb{C}}(p_iq_jA'p_iq_j \cap p_iq_jBp_iq_j)]^{\frac{1}{2}},$$

while

$$\tilde{\Lambda}_{ji} = [\dim_{\mathbb{C}}(q_j\tilde{p}_iB'q_j\tilde{p}_i \cap q_j\tilde{p}_iB_1q_j\tilde{p}_i)]^{\frac{1}{2}}.$$

Notice, by (ii) and (iii) above, that $q_j\tilde{p}_i = \tilde{p}_iq_j = Jp_iq_jJ$, so that

$$\begin{aligned} q_j\tilde{p}_iB'q_j\tilde{p}_i \cap q_j\tilde{p}_iB_1q_j\tilde{p}_i &= Jp_iq_jJB'Jp_iq_jJ \cap Jp_iq_jJB_1Jp_iq_jJ \\ &= J(p_iq_jBp_iq_j \cap p_iq_jA'p_iq_j)J, \end{aligned}$$

thereby showing that indeed, $\tilde{\Lambda}_{ji} = \Lambda_{ij}$.

(b) (i) Clearly the assignment $a \mapsto ae$ defines an injective *-homomorphism of A into eB_1e. To see that it is surjective, notice first that – thanks to the fundamental equation $ebe = E(b)e \ \forall b \in B$, where E denotes the ϕ-preserving conditional expectation of B onto A – the set $\{b_0 + \sum_{i=1}^n b_i e b_i' : n \geq 0, b_i, b_i' \in B\}$ is a *-subalgebra of B_1 which contains $B \cup \{e\}$ and is consequently equal to all of B_1; since $ebe = E(b)e, e(beb')e = E(b)E(b')e$, it follows that indeed $eB_1e = Ae$.

(ii) This follows at once from (b)(i), (a)(iii) and (i). □

We thus find that the isomorphism-type of $\langle B, e_A \rangle$ is independent of the initial state σ. It turns out, however, that there is one 'good choice' of trace on the big algebra – which is unique, under mild 'irreducibility' assumptions – as shown by the next result.

A word about notation before we state the next result (which is necessary because traces on finite-dimensional algebras are far from unique): if $A \subseteq B$ are as above, and if σ is a tracial state on B, we shall write $\sigma = \mathrm{tr}_{\overline{s}}$ if σ and $\overline{s}$ are related as in equation (3.2.3); further, in this case, we shall write $E_A^{\overline{s}}$ for the unique $\mathrm{tr}_{\overline{s}}$-preserving conditional expectation of B onto A.

PROPOSITION 3.2.3 *Let $\Lambda = \Lambda_A^B$ denote the inclusion matrix, where $A \subseteq B$ is an inclusion of finite-dimensional C^*-algebras as above; and let $\tau = \mathrm{tr}_{\overline{t}}$ be a tracial state on B, as described above. The following conditions on the trace τ are equivalent:*

(i) τ extends to a tracial state $\tau_1 = \mathrm{tr}_{\overline{t}_1}$ on $\langle B, e_A \rangle$ such that $E_B^{\overline{t}_1}(e_A) = \lambda 1$ for some scalar λ;

(ii) $\Lambda' \Lambda \overline{t} = \lambda^{-1} \overline{t}$.

Further, when these conditions are satisfied, the scalar λ must be the reciprocal of the Perron–Frobenius eigenvalue of the (positive semi-definite and entry-wise non-negative) matrix $\Lambda' \Lambda$, and hence,

$$\lambda^{-1} = ||\Lambda' \Lambda|| = ||\Lambda||^2.$$

Proof: Let $p_i, q_j, \tilde{p}_i$ be as in Lemma 3.2.2(a)(iv); we assume that it is these ordered sets of minimal projections in the three algebras A, B and $\langle B, e_A \rangle$ with respect to which inclusion matrices and trace-vectors are described. Thus, the inclusion matrix for $B \subseteq \langle B, e_A \rangle$ is just Λ' (by Lemma 3.2.2(a)(iv)).

Let $\tau_1 = \mathrm{tr}_{\overline{t}_1}$ be any tracial state on $\langle B, e_A \rangle$ which extends τ (or equivalently, $\overline{t}_1$ is a trace vector for $\langle B, e_A \rangle$ such that $\Lambda \overline{t}_1 = \overline{t}$). For each i, pick a minimal projection p_i^0 in A such that $p_i^0 \leq p_i$, and set $\widetilde{p_i^0} = p_i^0 e$. It follows from Lemma 3.2.2(b)(ii) that

$$(\Lambda \overline{t})_i = \tau(p_i^0) \quad \text{and} \quad (\overline{t}_1)_i = \tau_1(\widetilde{p_i^0}) \; \forall i. \tag{3.2.4}$$

On the other hand, it follows, by reasoning exactly as in the proof of Proposition 3.1.2(iv)(c), that

$$E_B^{\overline{t}_1}(e_A) = \lambda 1 \Leftrightarrow E_A^{\overline{t}_1}(e_A) = \lambda 1. \tag{3.2.5}$$

Hence condition (i) of the proposition is seen to be exactly equivalent to the requirement

(i)′ There exists a trace-vector $\bar{t}_1$ for $\langle B, e_A \rangle$ such that $\Lambda'\bar{t}_1 = \bar{t}$ and $\lambda^{-1}\bar{t}_1 = \Lambda\bar{t}$.

If (i)′ is satisfied, then $\Lambda'\Lambda\bar{t} = \lambda^{-1}\Lambda'\bar{t}_1 = \lambda^{-1}\bar{t}$ and (ii) is satisfied. Conversely, if (ii) is satisfied, and if we define $\bar{t}_1 = \lambda\Lambda\bar{t}$, we see that (ii)′ is satisfied.

DEFINITION 3.2.4 *A trace τ which satisfies the equivalent conditions of Proposition 3.2.3 is said to be a Markov trace for the inclusion $A \subseteq B$.*

COROLLARY 3.2.5 *Let $A \subseteq B$ be an inclusion of finite-dimensional C^*-algebras.*

(a) If τ is a Markov trace for the inclusion $A \subseteq B$, then it extends uniquely to a state $\mathrm{tr}_{\bar{t}_1}$ on $\langle B, e_A \rangle$ which is a Markov trace for the inclusion $B \subseteq \langle B, e_A \rangle$.

(b) The following conditions are equivalent:

(i) there exists a unique Markov trace for the inclusion $A \subseteq B$;

(ii) the Bratteli diagram for the inclusion $A \subseteq B$ is connected.

Proof: (a) This is clear from (the proof of) Proposition 3.2.3.

(b) The content of Proposition 3.2.3 is that $\mathrm{tr}_{\bar{t}}$ is a Markov trace for the inclusion $A \subseteq B$ precisely when $\bar{t}$ is a Perron–Frobenius eigenvector of the (entry-wise non-negative) matrix $\Lambda\Lambda'$. The point is that the general theory says – see [Gant], for instance – when such an eigenvector is unique, and that answer, when translated into our context, is that (ii) is precisely what is needed to ensure that uniqueness. □

3.3 The projections e_n and the tower

In this section, we shall be interested in an (initial) inclusion $M_{-1} \subseteq M_0$ of finite von Neumann algebras, which falls into one of the following cases:

Case (i) M_0 and M_{-1} are II_1 factors, and $[M_0 : M_{-1}] < \infty$;

Case (ii) M_0 and M_{-1} are finite-dimensional; in this case, we shall always assume that the inclusion is 'connected' (meaning that the Bratteli diagram is connected), and we shall reserve the symbol tr_{M_0} for the Markov trace for the inclusion, which is unique by Corollary 3.2.5(b).

In either case, write $M_1 = \langle M_0, e_1 \rangle$, where we write e_1 for the projection we denoted earlier by $e_{M_{-1}}$. The reason for the changed notation is that then the inclusion $M_0 \subseteq M_1$ falls into the same 'case' above, as did the initial inclusion $M_{-1} \subseteq M_0$; hence, we can iterate the above procedure to obtain a tower

$$M_1 \subseteq M_2 \subseteq \cdots \subseteq M_n \subseteq M_{n+1} \subseteq \cdots \tag{3.3.1}$$

where $M_{n+1} = \langle M_n, e_{n+1} \rangle$ is the result of applying the basic construction to the inclusion $M_{n-1} \subseteq M_n$, and e_{n+1} denotes the projection implementing the (tr_{M_n}-preserving) conditional expectation of M_n onto M_{n-1}.

It must be clear that in case (i), the tower (3.3.1) is a tower of II_1 factors such that $[M_{n+1} : M_n]$ is independent of n; and that in case (ii), it is a tower of finite-dimensional C^*-algebras. In either case, we see that $\bigcup M_n$ comes equipped with a tracial state tr (whose restriction to M_n is tr_{M_n}); in fact this is the unique tracial state on $\bigcup M_n$, and consequently we see that $M_\infty = \pi_{\mathrm{tr}}(\bigcup M_n)''$ is a II_1 factor (which is hyperfinite in case (ii), and also in case (i) provided that each of M_0 and M_{-1} is).

EXAMPLE 3.3.1 *(i) Suppose* $M_{-1} = \mathbb{C}1 \subseteq M_N(\mathbb{C}) = M_0$; *then the inclusion matrix* $\Lambda_{M_{-1}}^{M_0}$ *is the* 1×1 *matrix* $[N]$. *It follows that* $M_{n-1} \cong M_{N^n}(\mathbb{C}) \cong \bigotimes^n M_N(\mathbb{C})\ \forall n \geq 0$, *and we find that* $M_\infty = R_{(N)}$. *If we identify* M_0 *with the subalgebra* $M_N(\mathbb{C}) \otimes 1$ *of* $M_N(\mathbb{C}) \otimes M_N(\mathbb{C}) = M_1$, *then the projection* e_1 *is given, in terms of the usual system* $\{e_{ij}\}_{i,j=1}^N$ *of matrix units of* $M_N(\mathbb{C})$, *by the formula* $e_1 = \frac{1}{N}\sum_{i,j=1}^N e_{ij} \otimes e_{ij}$.

(ii) (In a sense, this is a 'square root' of the last example.)

Suppose $M_{-1} = \mathbb{C}1 \subseteq \mathbb{C}^N = M_0$. *Then* $\Lambda_{M_{-1}}^{M_0} = [1\ 1 \cdots 1]$ *and it follows that* $M_{2n-1} \cong M_{N^n}(\mathbb{C})\ \forall n \geq 1$. *If we think of* M_0 *as the diagonal subalgebra of* $M_N(\mathbb{C}) = M_1$, *then the projection* e_1 *is the* $N \times N$ *matrix with all entries equal to* $\frac{1}{N}$.

The sequence $\{e_n\}_{n=1}^\infty$ of projections plays a central role in the theory of subfactors. We list below some properties of this sequence.

PROPOSITION 3.3.2 *Let* $\{e_n\}_{n=1}^\infty$ *be the sequence of projections in the* II_1 *factor* M_∞, *constructed, as above, from an initial inclusion* $M_{-1} \subseteq M_0$ *which falls into either case(i) or case (ii). Write* tr *for the unique tracial state on* M_∞. *Then:*

(i) the number $\tau = \mathrm{tr}\, e_n$ *is independent of* n; *further,*

$$\tau^{-1} = \begin{cases} [M_0 : M_{-1}] & \text{in case (i),} \\ ||\Lambda_{M_{-1}}^{M_0}||^2 & \text{in case (ii);} \end{cases}$$

(ii) $\mathrm{tr}(xe_n) = \tau\, \mathrm{tr}\, x\ \forall x \in M_{n-1}, n \geq 1$;

(iii) $e_n \in M_{n-2}' \cap M_n\ \forall n \geq 1$, *and in particular,*

$$e_n e_m = e_m e_n \quad \text{if } |m - n| > 1;$$

(iv) $e_{n+1}e_ne_{n+1} = \tau e_{n+1}\ \forall n \geq 1$;

(v) $e_ne_{n+1}e_n = \tau e_n\ \forall n \geq 1$.

Proof: The Markov property and the fact that e_n has trace equal to τ are proved in Proposition 3.1.2 (for case (i)) and Proposition 3.2.3 (for case (ii)); this takes care of (ii) and (i).

Assertion (iii) is a consequence of Proposition 3.1.2 (i).

(iv) Since tr restricts, on M_n, to tr_{M_n}, it follows from equation (3.1.1) that $e_{n+1}xe_{n+1} = E_{M_{n-1}}(x)e_{n+1}\ \forall x \in M_n$; in particular, if we choose $x = e_n$ and use the already proved (ii), we get (iv).

(v) It follows from (iv) that $u = \tau^{-\frac{1}{2}}e_ne_{n+1}$ is a partial isometry in M_∞, with initial projection given by $u^*u = e_{n+1}$. The final projection of u clearly satisfies $uu^* \leq e_n$. On the other hand, we must have

$$\tau = \mathrm{tr}\, e_n \geq \mathrm{tr}(uu^*) = \mathrm{tr}(u^*u) = \mathrm{tr}\, e_{n+1} = \tau;$$

thus we must have $\mathrm{tr}\, uu^* = \mathrm{tr}\, e_n$; since tr is a faithful trace, this implies that $uu^* = e_n$, as desired. □

Notice the two expressions for τ in Proposition 3.3.2(i); this is just an indication of the deeper relationship between index of subfactors and squares of norms of non-negative integer matrices. Parallel to Theorem 2.3.8, there is an old result which is essentially due to Kronecker – see [GHJ] – which says that if Λ is a finite matrix with integral entries, then

$$||\Lambda|| \in \{2\cos\frac{\pi}{n} : n = 2, 3, 4, \cdots\} \cup [2, \infty].$$

To better understand this statement, recall that – exactly as the inclusion matrix Λ_A^B is related to the Bratteli diagram for the inclusion $A \subseteq B$ of finite-dimensional C^*-algebras – there is a bijective correspondence between finite matrices with non-negative integral entries and finite bipartite (multi-) graphs. The matrix G which corresponds to a bipartite graph $\mathcal{G}$ is referred to as the 'adjacency matrix of $\mathcal{G}$' and $||G||$ will also be called the norm of the graph $\mathcal{G}$. Closely tied up with Kronecker's theorem (and the well-known classification of Coxeter graphs or Dynkin diagrams) is the fact that the only (bipartite) graphs with norm less than two are the ones listed below, where the subscript refers to the number of vertices in the graph.

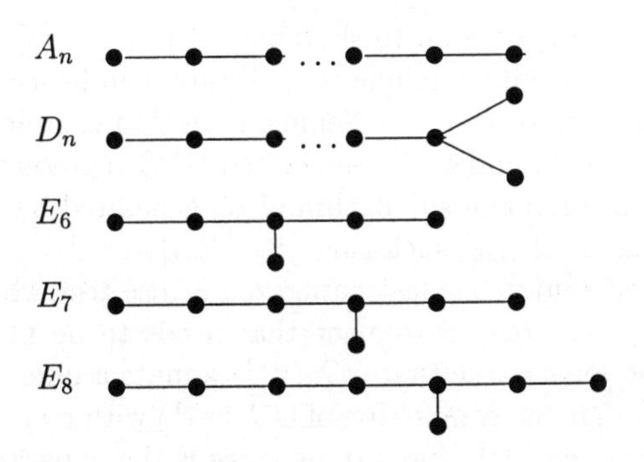

Now return to the case of a general initial inclusion $M_{-1} \subseteq M_0$ (which falls into case (i) or case (ii)). Let e_n, τ be as in Proposition 3.3.2(i). (Thus, according to the foregoing discussion and Theorem 2.3.8, either $\tau^{-1} \geq 4$ or $\tau^{-1} = 4\cos^2 \frac{\pi}{m}$ for some integer $m \geq 3$.)

Consider the algebras defined by

$$Q_n(\tau^{-1}) = \begin{cases} \text{Alg}\{e_1, \cdots, e_n\} = \{e_1, \cdots, e_n\}'' & \text{if } n > 0, \\ \mathbb{C} & \text{if } n \in \{-1, 0\}. \end{cases}$$

(This notation turns out to be justified, as the following discussion shows.)

It is an easy consequence of the properties in Proposition 3.3.2(iii), (iv) and (v) that each $Q_n(\tau^{-1})$ is finite-dimensional. It turns out – see [Jon1] for details – that for any $\tau \leq \frac{1}{4}$, the Bratteli diagram for the tower $\{Q_n(\tau^{-1})\}_{n=-1}^{\infty}$ is given by a 'half-Pascal-triangle', and that the Bratteli diagram for the tower $\{Q_n(4\cos^2 \frac{\pi}{m})\}_{n=-1}^{\infty}$ is obtained by starting with the half-Pascal-triangle and slicing it off in a vertical line between the vertices labelled $(m-3)$ and $(m-2)$. (The cases $m = 4, 5$ have been illustrated below.)

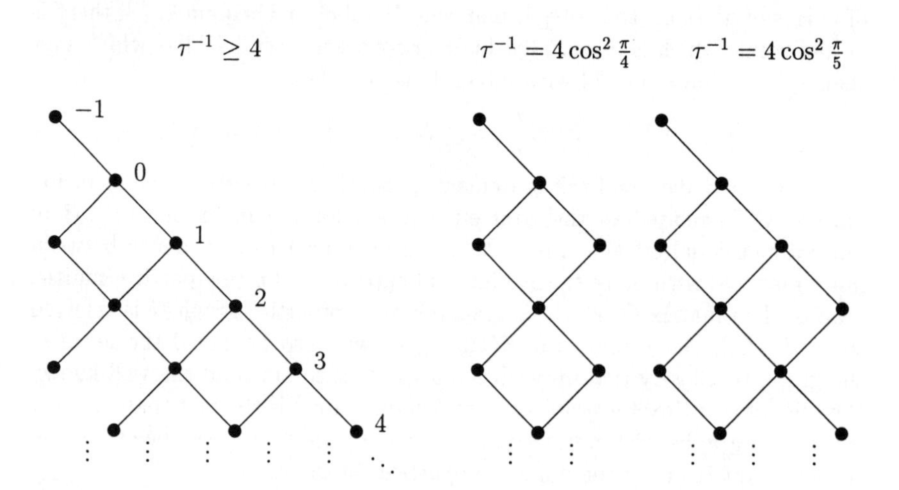

The analysis in [Jon1] goes on to show that if $\tau = \frac{1}{4} sec^2 \frac{\pi}{m}$ where $m \geq 3$, the algebra $\cup Q_n(\tau^{-1})$ admits a unique tracial state, and hence has a copy R_0 of the hyperfinite II_1 factor as a von Neumann algebra completion (meaning weak closure of its image under the associated GNS representation); and if R_{-1} denotes the von Neumann subalgebra of R_0 generated by $\{e_n : n \geq 2\}$, then R_{-1} is a subfactor of R_0 such that $[R_0 : R_{-1}] = \tau^{-1}$.

Most of what was said in the last paragraph is also true when $\tau \leq 2$ (and is proved in [Jon1]); the only statement that needs to be modified is that when $\tau \leq 2$, it is no longer true that $\cup Q_n(\tau^{-1})$ admits a unique trace, but it is true nevertheless that the completion of $\cup Q_n(\tau^{-1})$ with respect to the GNS representation associated with the Markov trace is the hyperfinite factor R.

Hence, in order to exhibit a subfactor of R with index $4\cos^2\frac{\pi}{m}$, it suffices to find an inclusion $M_{-1} \subseteq M_0$ whose Bratteli diagram has norm $2\cos\frac{\pi}{m}$ – and for this, we may choose the graph A_{m-1} and pick an inclusion with this as Bratteli diagram – and then $R_{-1} \subseteq R_0$ meets our requirements.

Before concluding this section, we wish to point out that if τ, R_{-1}, R_0 are as above, we have an instance of simultaneous approximation of the subfactor–factor inclusion $R_{-1} \subseteq R_0$ by finite-dimensional inclusions in the following sense :

$$\begin{array}{ccccccccccc}
\langle e_1 \rangle & \subseteq & \langle e_1, e_2 \rangle & \subseteq & \cdots & \subseteq & \langle e_1, e_2, \cdots, e_n \rangle & \subseteq & \cdots & \rightarrow & R_0 \\
\cup & & \cup & & & & \cup & & & & \cup \\
\mathbb{C} & \subseteq & \langle e_2 \rangle & \subseteq & \cdots & \subseteq & \langle e_2, \cdots, e_n \rangle & \subseteq & \cdots & \rightarrow & R_{-1}
\end{array}$$

Chapter 4

The principal and dual graphs

4.1 More on bimodules

Suppose M, P are arbitrary von Neumann algebras with separable pre-duals, and suppose $\mathcal{H}$ is a separable M-P-bimodule. Pick some faithful normal state φ and set $\mathcal{H}_1 = L^2(M, \varphi)$ and $\mathcal{H}_\infty = \mathcal{H}_1 \otimes \ell^2$. It follows from Theorem 2.2.2 that $\mathcal{H}$ may be identified, as a left M-module, with $\mathcal{H}_\infty q$ for some projection $q \in M_\infty(M)$ (which is uniquely determined up to Murray–von Neumann equivalence in $M_\infty(M)$); further, we have ${}_M\mathcal{L}(\mathcal{H}) = \pi_r(M_\infty(M)_q)$. Since $\mathcal{H}$ is an M-P-bimodule, it follows from our identification that there exists a normal unital homomorphism $\theta : P \to M_\infty(M)_q$ such that the right action of P is given by $\xi \cdot y = \xi\theta(y)$.

Conversely, given a normal homomorphism $\theta : P \to M_\infty(M)$, let $\mathcal{H}_\theta$ denote the M-P-bimodule with underlying Hilbert space $\mathcal{H}_\infty\theta(1)$, and with the actions given, via matrix multiplication, by $m \cdot \xi \cdot p = m\xi\theta(p)$. The content of the preceding paragraph is that every separable M-P-bimodule is isomorphic to $\mathcal{H}_\theta$ for suitable θ.

If M is a factor of type III, then so is $M_\infty(M)$, and hence every non-zero projection in $M_\infty(M)$ is Murray–von Neumann equivalent to $1_M \otimes e_{11}$. Consequently, every M-M-bimodule is isomorphic to $\mathcal{H}_\theta$ for some endomorphism $\theta : M \to M$.

Suppose M and P are II_1 factors and suppose $\mathcal{H}_\theta$ is as above. (In this case, we naturally take $\varphi = \mathrm{tr}_M$.) Notice then that $\dim_{M-}\mathcal{H}$ is finite precisely when $\mathrm{Tr}\,\theta(1) < \infty$, while $\dim_{-P}\mathcal{H}$ is finite precisely when the index $[M_n(M)_q : \theta(P)]$ is finite.

This suggests that we define a *co-finite morphism of P into M* as a normal homomorphism $\theta : P \to M_\infty(M)$ such that (i) $\theta(1)$ is a finite projection, say q, in $M_\infty(M)$, and (ii) $\theta(P)$ has finite index in $M_\infty(M)_q$. The point is that if θ is a co-finite morphism of P into M, then $\mathcal{H}_\theta = \mathcal{H}_\infty\theta(1)$ is a bifinite M-P-bimodule, and every bifinite M-P-bimodule arises in this fashion.

Motivated by the case of automorphisms, we shall say that two co-finite morphisms $\theta_i, i = 1, 2$, of P into M are outer equivalent if there exists a partial isometry $u \in M_\infty(M)$ such that $u^*u = \theta_1(1), uu^* = \theta_2(1)$ and $\theta_2(x) = u\theta_1(x)u^* \ \forall x \in P$; this is because, with this definition, two co-finite morphisms θ_i from P into M are outer equivalent if and only if the corresponding bifinite M-P-bimodules $\mathcal{H}_{\theta_i}$ are isomorphic.

We wish, in the rest of this section, to discuss two operations – one unary (contragredient) and one binary (tensor products) – that can be performed with bimodules. (In the latter case, we shall restrict ourselves to the case when both the algebras in question are II_1 factors.)

Contragredients: Suppose $\mathcal{H}$ is an M-P-bimodule. By the contragredient of $\mathcal{H}$, we shall mean any P-M-bimodule $\overline{\mathcal{H}}$ for which there exists an anti- unitary operator $J : \mathcal{H} \to \overline{\mathcal{H}}$ such that $J(m \cdot \xi \cdot p) = p^* \cdot J\xi \cdot m^*$. It is clear that such a contragredient exists and is unique up to isomorphism.

Note that if $\mathcal{H} = \mathcal{H}_\theta$, and if $\overline{\mathcal{H}} \cong \mathcal{H}_{\overline{\theta}}$, then $\theta : P \to M_\infty(M)$ while $\overline{\theta} : M \to M_\infty(P)$, and the relationship between the morphisms θ and $\overline{\theta}$ is somewhat mysterious. However, in the simple case when $M = P$ and θ is an automorphism of M, it is easy to verify that $\overline{\theta}$ (which is, after all, only determined up to outer equivalence) may be taken as θ^{-1}. (The reader should have no difficulty in constructing a J which establishes that $\mathcal{H}_{\theta^{-1}}$ is a contragedient of $\mathcal{H}_\theta$.)

Tensor products: We shall discuss tensor products of bifinite bimodules over II_1 factors. First, however, we want to single out a distinguished dense subspace of such a bimodule, namely the one consisting of the so-called bounded vectors.

Suppose, to be specific, that M, P are II_1 factors and that $\mathcal{H}$ is a bifinite M-P-bimodule. Say that a vector $\xi \in \mathcal{H}$ is *left-bounded* (resp., *right-bounded*) if there exists a constant $K > 0$ such that $||\xi p||^2 \leq K\text{tr}_P(p^*p) \ \ \forall p \in P$ (resp., $||m\xi||^2 \leq K\text{tr}_M(m^*m) \ \ \forall m \in M$), or equivalently, if there exists a bounded operator $L_\xi : L^2(P) \to \mathcal{H}$ (resp., $R_\xi : L^2(M) \to \mathcal{H}$) such that $L_\xi(p\Omega) = \xi p \ \ \forall p \in P$ (resp., $R_\xi(m\Omega) = m\xi \ \ \forall m \in M$). It is true that a vector is left-bounded if and only if it is right-bounded. (Reason: assume $\mathcal{H} = \mathcal{H}_\theta$, for a co-finite morphism $\theta : P \to M_n(M)$; thus a vector in $\mathcal{H}$ is of the form $\xi = (\xi_1, \cdots, \xi_n) \in (\text{Mat}_{1\times n}(L^2(M))\theta(1)$; it follows easily from Theorem 1.2.4(1) – and the fairly easily proved fact that if P_0 is a subfactor of P of finite index, then a vector is left-bounded for the right action of P if and only if it is left-bounded for the right action of P_0 – that the above vector ξ satisfies either of the boundedness conditions above precisely when each co-ordinate has the form $\xi_i = x_i\Omega$ for some $x_i \in M$.) Thus we may talk simply of *bounded vectors*. We shall denote the collection of bounded vectors in the bifinite bimodule $\mathcal{H}$ by the symbol $\mathcal{H}_0$.

We list some properties of the assignment $\mathcal{H} \mapsto \mathcal{H}_0$ in the following proposition, which is easily proved by considering the case of the model $\mathcal{H}_\theta$. (In any case, complete proofs of all the assertions in this section may be found, for instance, in [Sun3]; actually, only the case $M = P$ is treated there, but the proofs carry over verbatim to the 'more general' case.)

PROPOSITION 4.1.1 *(a) Let M, P be II_1 factors and let $\mathcal{H}, \mathcal{K}$ denote arbitrary bifinite M-P-bimodules. Then $\mathcal{H}_0$ is an M-P-linear dense subspace of $\mathcal{H}$, and the assignment $T \mapsto T|_{\mathcal{H}_0}$ defines a bijective correspondence between the Banach space ${}_M\mathcal{L}_P(\mathcal{H}, \mathcal{K})$ and the vector space ${}_ML_P(\mathcal{H}_0, \mathcal{K}_0)$ of M-P-linear transformations between the vector spaces $\mathcal{H}_0$ and $\mathcal{K}_0$.*

(b) There exists a unique mapping $(\xi, \eta) \mapsto \langle \xi, \eta \rangle_M$ from $\mathcal{H}_0 \times \mathcal{H}_0$ into M, referred to as the M-valued inner product on the bimodule $\mathcal{H}$, *which satisfies the following properties, for all $\xi, \eta, \zeta \in \mathcal{H}_0, m \in M, p \in P$:*

(i) $\langle \xi, \eta \rangle = \mathrm{tr}_M(\langle \xi, \eta \rangle_M)$.

(ii) $\langle \xi, \xi \rangle_M \geq 0$.

(iii) $\langle \xi, \eta \rangle_M = (\langle \eta, \xi \rangle_M)^*$.

(iv) $\langle m \cdot \xi + \zeta, \eta \rangle_M = m(\langle \xi, \eta \rangle_M) + \langle \zeta, \eta \rangle_M$.

(v) $\langle \xi \cdot p, \eta \rangle_M = \langle \xi, \eta \cdot p^* \rangle_M$. □

As has been already remarked, if $\mathcal{H} = \mathcal{H}_\theta$, then we may identify $\mathcal{H}_0$ with $\mathrm{Mat}_{1\times n}(M)\theta(1)$, and in this case,

$$\langle (x_1, \cdots, x_n), (y_1, \cdots, y_n) \rangle_M = \sum_i x_i y_i^*.$$

Further, if $\mathcal{K} = \mathcal{H}_\phi$, then the typical element of ${}_M\mathcal{L}_P(\mathcal{H}, \mathcal{K})$ is of the form $\xi \mapsto \xi T$ (matrix multiplication), where $T \in \theta(1)M_\infty(M)\phi(1)$ is a matrix satisfying $\theta(p)T = T\phi(p)$ $\forall p \in P$.

One reason for introducing the M-valued inner product is to facilitate the formulation of the universal property possessed by the tensor product.

PROPOSITION 4.1.2 *Suppose M, P and Q are II_1 factors, and suppose $\mathcal{H}$ (resp., $\mathcal{K}$) is a bifinite M-P-bimodule (resp., P-Q-bimodule). Then there exists a bifinite M-Q-bimodule, denoted by $\mathcal{H} \otimes_P \mathcal{K}$, which is determined uniquely up to isomorphism, by the following property:*

There exists a surjective linear map from the algebraic tensor product $\mathcal{H}_0 \otimes \mathcal{K}_0$ onto $(\mathcal{H} \otimes_P \mathcal{K})_0$, the image of $\xi \otimes \eta$ being denoted by $\xi \otimes_P \eta$, satisfying:

(a) $\xi \cdot p \otimes_P \eta = \xi \otimes_P p \cdot \eta$;

(b) $m \cdot \xi \otimes_P \eta \cdot q = m \cdot (\xi \otimes_P \eta) \cdot q$;

(c) $\langle \xi \otimes_P \eta, \xi' \otimes_P \eta' \rangle_M = \langle \xi \cdot \langle \eta, \eta' \rangle_P, \xi' \rangle_M$. □

The useful way to think of these tensor products is in the reformulation in terms of co-finite morphisms. Suppose, to be specific, that $\mathcal{H} = \mathcal{H}_\theta, \mathcal{K} = \mathcal{H}_\phi$ for some co-finite morphisms $\theta : P \to M_m(M), \phi : Q \to M_n(P)$. It is not hard to see that the equation

$$(\theta \otimes \phi)_{ij,kl}(q) = \theta_{ik}(\phi_{jl}(q)) \tag{4.1.1}$$

defines a normal homomorphism $\theta \otimes \phi : Q \to M_{mn}(M)$; it is further true (although co-finiteness requires some work) that $\theta \otimes \phi$ is a co-finite morphism from Q into M and that $\mathcal{H}_{\theta\otimes\phi} \cong \mathcal{H}_\theta \otimes_P \mathcal{H}_\phi$. In particular, in the special case when $M = P = Q$ and $\theta, \phi \in \mathrm{Aut}(M)$, we find the reassuring fact that $\theta \otimes \phi = \theta \circ \phi$.

Another immediate consequence of the morphism description of the tensor product is the multiplicativity of dimension under tensor products, meaning that if $\mathcal{H}$ (resp. $\mathcal{K}$) is an M-N- (resp., N-P-) bimodule, then

$$\dim_{M-}(\mathcal{H} \otimes_N \mathcal{K}) = \dim_{M-} \mathcal{H} \cdot \dim_{N-} \mathcal{K} \tag{4.1.2}$$

and

$$\dim_{-P}(\mathcal{H} \otimes_N \mathcal{K}) = \dim_{-N} \mathcal{H} \cdot \dim_{-P} \mathcal{K}. \tag{4.1.3}$$

4.2 The principal graphs

We assume throughout this section that $N \subseteq M$ is an inclusion of II_1 factors such that $[M : N] < \infty$, and that

$$N = M_{-1} \subseteq M = M_0 \subseteq \cdots \subseteq M_1 \subseteq \cdots \subseteq M_n \subseteq M_{n+1} \subseteq \cdots \tag{4.2.1}$$

is the tower of the basic construction, with $M_{n+1} = \langle M_n, e_{n+1}\rangle$ for $n \geq 0$.

It follows from Proposition 2.3.12 that $\{M_i' \cap M_j : -1 \leq i \leq j\}$ is a grid of finite-dimensional C^*-algebras, which is canonically associated with the inclusion $N \subseteq M$, and is consequently an 'invariant' of the initial inclusion. It turns out – see Proposition 4.3.7 – that there is a periodicity of order two and hence we need to consider only $i = -1$ and $i = 0$.

Let us consider $i = -1$ first. Notice that e_{n+1} belongs to $N' \cap M_{n+1}$ and implements the conditional expectation of $N' \cap M_n$ onto $N' \cap M_{n-1}$. It follows – from Lemma 5.3.1(b) – that $N' \cap M_{n+1}$ contains a copy of the basic construction for the inclusion $(N' \cap M_{n-1}) \subseteq (N' \cap M_n)$, and consequently the Bratteli diagram for the inclusion $(N' \cap M_n) \subseteq (N' \cap M_{n+1})$ contains a 'reflection' of the Bratteli diagram for the inclusion $(N' \cap M_{n-1}) \subseteq (N' \cap M_n)$. The graph obtained by starting with the Bratteli diagram for the tower $\{N' \cap M_n : n \geq -1\}$ of relative commutants, and removing all those parts which are obtained by reflecting the previous stage, is called the *principal graph* invariant of the inclusion $N \subseteq M$. A similar reasoning also applies for $i = 0$, and the resulting graph is called the *dual graph* invariant of the inclusion $N \subseteq M$.

We shall now give an alternative description of these two graphs, using the language of bimodules. (The equivalence of these two descriptions is established in §4.4.)

The principal graph is the bipartite multi-graph $\mathcal{G}$ defined as follows: the set $\mathcal{G}^{(0)}$ of even vertices is indexed by isomorphism classes of irreducible N-N-bimodules which occur as submodules of ${}_N L^2(M_n)_N$ for some $n \geq 1$; the set $\mathcal{G}^{(1)}$ of odd vertices is indexed by isomorphism classes of irreducible N-M-bimodules which occur as submodules of ${}_N L^2(M_n)_M$ for some $n \geq 1$; the even vertex labelled by the irreducible N-N-bimodule X is connected to the odd vertex labelled by the irreducible N-M-bimodule Y by k bonds, if k is the multiplicity with which X occurs in the N-N-bimodule Y.

The dual graph is the bipartite multi-graph $\mathcal{H}$ defined as follows: the set $\mathcal{H}^{(0)}$ of even vertices is indexed by isomorphism classes of irreducible M-M-bimodules which occur as submodules of ${}_M L^2(M_n)_M$ for some $n \geq 1$; the set $\mathcal{H}^{(1)}$ of odd vertices is indexed by isomorphism classes of irreducible M-N-bimodules which occur as submodules of ${}_M L^2(M_n)_N$ for some $n \geq 1$; the even vertex labelled by the irreducible M-M-bimodule X is connected to the odd vertex labelled by the irreducible M-N-bimodule Y by k bonds, if k is the multiplicity with which Y occurs in the M-N-bimodule X.

It is a consequence of Proposition 4.3.7 that the principal (resp., dual) graph for the inclusion $M \subseteq M_1$ may be identified with the dual (resp., principal) graph for the inclusion $N \subseteq M$. Hence any result concerning the principal graph has a corresponding statement about the dual graph. So we shall restrict ourselves, in this section, to making some comments concerning the principal graph.

The principal graph has a distinguished vertex $*_{\mathcal{G}}$ which corresponds to the even vertex indexed by the isomorphism class of the standard bimodule ${}_N L^2(N)_N$. Notice that the odd vertices at distance 1 from $*_{\mathcal{G}}$ are indexed by isomorphism classes of irreducible N-M-submodules of ${}_N L^2(M)_M$; hence the case of irreducible subfactors corresponds to the case where $*_{\mathcal{G}}$ has a unique neighbour. Further, since the tensor product of bifinite bimodules is a direct sum of only finitely many irreducible submodules, it follows from the description of the principal graph that each vertex in $\mathcal{G}$ has only finitely many edges incident on it; thus the principal graph is a *locally finite connected pointed bipartite graph.*

Suppose $\mathcal{G}$ is the principal graph of the inclusion $N \subseteq M$, as above. Let G denote the (non-negative integer-valued) $\mathcal{G}^{(0)} \times \mathcal{G}^{(1)}$ matrix with g_{XY} equal to the number of bonds joining the even vertex X to the odd vertex Y in the graph $\mathcal{G}$. We shall also sometimes use the suggestive notation $g_{XY} = \langle X, Y \rangle$. Even if the graph $\mathcal{G}$ is infinite, the local finiteness of $\mathcal{G}$ translates into the fact that the matrix G has only finitely many non-zero entries on any row or column.

We are now ready to prove an important relation between the index of the subfactor and the norm of the matrix G, which will, in particular, reduce the statement in Theorem 2.3.8 concerning the restriction on index values of subfactors to the classification of graphs of norm less than 2.

PROPOSITION 4.2.1 *Let $N \subset M$ be an inclusion of II_1 factors such that $[M:N] < \infty$, and let $G, \mathcal{G}$ be as above.*

(i) G defines, via matrix multiplication, a bounded operator from $\ell^2(\mathcal{G}^{(1)})$ into $\ell^2(\mathcal{G}^{(0)})$, with $||G||^2 \leq [M:N]$.

(ii) If $[M:N] < 4$, then $\mathcal{G}$ is necessarily finite and is one of the Coxeter diagrams A_n, D_n, E_6, E_7 or E_8.

Proof: (i) In order to prove (i), we shall prove the clearly equivalent statement that GG' defines a bounded self-adjoint operator on $\ell^2(\mathcal{G}^{(0)})$ with norm at most $[M:N]$.

It is a consequence of the 'Frobenius reciprocity' statement contained in Proposition 4.4.1 that

$$({}_N X_N) \otimes_N ({}_N L^2(M)_M) \cong \bigoplus_{Y \in \mathcal{G}^{(1)}} \langle X, Y \rangle \cdot {}_N Y_M, \quad \forall X \in \mathcal{G}^{(0)},$$

where we write $m \cdot \mathcal{H}$ to denote the direct sum of m copies of the module $\mathcal{H}$. Equating left and right dimensions, we get

$$[M:N] \dim_{N-} X = \sum_{Y \in \mathcal{G}^{(1)}} g_{XY} \dim_{N-} Y, \quad \forall X \in \mathcal{G}^{(0)}, \tag{4.2.2}$$

and

$$\dim_{-N} X = \sum_{Y \in \mathcal{G}^{(1)}} g_{XY} \dim_{-M} Y, \quad \forall X \in \mathcal{G}^{(0)}. \tag{4.2.3}$$

Similarly, we find that

$$({}_N Y_M) \otimes_M ({}_M L^2(M)_N) \cong \bigoplus_{X \in \mathcal{G}^{(0)}} \langle X, Y \rangle \cdot {}_N X_N, \quad \forall Y \in \mathcal{G}^{(1)}.$$

Equating dimensions, we get

$$\dim_{N-} Y = \sum_{X \in \mathcal{G}^{(0)}} g_{XY} \dim_{N-} X, \quad \forall Y \in \mathcal{G}^{(1)}, \tag{4.2.4}$$

and

$$[M:N] \dim_{-M} Y = \sum_{X \in \mathcal{G}^{(0)}} g_{XY} \dim_{-N} X, \quad \forall Y \in \mathcal{G}^{(1)}. \tag{4.2.5}$$

It is an immediate consequence of equations (4.2.2) and (4.2.4) that if we define $v_X = \dim_{N-} X, \ \ \forall X \in \mathcal{G}^{(0)}$, then v is a (column-) vector indexed by $\mathcal{G}^{(0)}$ such that $(GG')v = [M:N]v$.

On the other hand, it is a consequence of the Perron–Frobenius theorem – see [Gant], for instance – that if the adjacency matrix of a connected graph has a positive eigenvector, then that adjacency matrix defines a bounded operator (on the ℓ^2 space with basis indexed by the vertices of the graph) with norm bounded by the eigenvalue afforded by the positive vector. This completes the proof of (i).

(ii) If $[M : N] < 4$, it follows from (i) that $||G|| < 2$, and we may appeal to the classification of matrices of norm less than 2 – see the discussion in §3.3. □

We now discuss some examples of principal graphs.

EXAMPLE 4.2.2 *Let $N = \langle 1, e_2, e_3, \cdots \rangle \subseteq M = \langle 1, e_1, e_2, \cdots \rangle$, as in §3.3. It is a consequence of a result called Skau's lemma – see [GHJ] for details – that $\mathcal{G} = A_{n-1}$ if $[M : N] = 4\cos^2 \frac{\pi}{n}$; thus, in these cases, we find that the Bratteli diagram for the tower $\{N' \cap M_n : n \geq -1\}$ is precisely the truncated Pascal triangle of §3.3. (Since $N' \cap M_n \supseteq \langle 1, e_1, \cdots, e_n \rangle$, this means, in view of the description of the Bratteli diagram for the tower $\{Q_n(\tau^{-1})\}$ discussed in §3.3, that the inclusion above is actually an equality.)*

When $[M : N] > 4$, however, things are quite different – see [GHJ] – and the principal graph turns out to be $A_{-\infty,\infty}$ which is the infinite path extending to infinity in both directions, so that we have a strict inclusion $N' \cap M_n \supset \langle 1, e_1, \cdots, e_n \rangle$ in this case, and the Bratteli diagram for the tower $\{N' \cap M_n : n \geq -1\}$ turns out to be the full Pascal triangle.

EXAMPLE 4.2.3 *Suppose G is a finite group acting as outer automorphisms of a II_1 factor P.*

(i) Let $N = P \subseteq P \times G = M$. In this case, the principal graph is an n-star (where $n = |G|$) with all arms of length one; thus, for instance, if $|G| = 3$, the principal graph is just the Coxeter graph D_4.

(ii) Let $N = P^G \subseteq P = M$ be the subfactor of fixed points under the action. Then the principal graph $\mathcal{G}$ has one odd vertex and the even vertices are indexed by the inequivalent irreducible representations of G, with the even vertex indexed by π connected to the unique odd vertex by d_π bonds, where d_π denotes the degree of the representation π.

It is a fact – see §A.4 – that if $M_1 = \langle M, e_N \rangle$ is the result of the basic construction, then the inclusion $M \subseteq M_1$ is isomorphic to the inclusion $M \subseteq M \times G$, and hence the discussion in the previous paragraph amounts to a description of the dual graph of the case considered in (i).

(iii) More generally than in (i) and (ii), if H is a subgroup of G, we could consider the case $N = P \times H \subseteq P \times G = M$. This case is treated in §A.4, where the principal and dual graphs are explicitly computed, using the bimodule approach.

EXAMPLE 4.2.4 *Suppose $\{\theta_1, \cdots, \theta_n\}$ is a set of automorphisms of R, which is closed under the formation of inverses. It has then been shown by Popa –*

see [Pop6] – that if

$$N = \left\{ \begin{bmatrix} x & 0 & \cdots & 0 \\ 0 & \theta_1(x) & \cdots & 0 \\ \vdots & \vdots & \ddots & \vdots \\ 0 & 0 & \cdots & \theta_n(x) \end{bmatrix} : x \in R \right\} \subseteq M_{n+1}(R) = M$$

denotes the so-called 'diagonal subfactor' (given by the θ_i's), then the principal graph is described by what might be called the Cayley graph of the subgroup of $\mathrm{Aut}(R)/\mathrm{Int}(R)$ *generated by the θ_i's – where* $\mathrm{Int}(R)$ *denotes the group of inner automorphisms of R.*

4.3 Bases

We assume in the rest of this chapter that $N \subseteq M$ is a finite-index inclusion of II_1 factors, and that

$$N = M_{-1} \subseteq M = M_0 \subseteq M_1 \subseteq \cdots \subseteq M_n \subseteq M_{n+1} \subseteq \cdots \tag{4.3.1}$$

is the tower of the basic construction, with $M_n = \langle M_{n-1}, e_n \rangle$ for $n \geq 1$. Also we use the notation $\tau = [M : N]^{-1}$. Further, we shall identify M with the dense subspace $M\Omega$ of $L^2(M)$.

We begin with a very useful fact.

LEMMA 4.3.1 *(i) If $x_1 \in M_1$, there exists a unique element $x_0 \in M$ such that $x_1 e_1 = x_0 e_1$; this element is given by $x_0 = \tau^{-1} E_M(x_1 e_1)$.*

(ii) The action of M_1 on $L^2(M)$ is given by

$$x_1(y_0 \Omega_M) = \tau^{-1} E_M(x_1 y_0 e_1) \Omega_M.$$

Proof: (i) Since $E_M(e_1) = \tau$, the second assertion and the uniqueness in the first assertion follow. As for existence, we need therefore to show that $\tau^{-1} E_M(x_1 e_1) e_1 = x_1 e_1$ for all $x_1 \in M_1$. Since the two sides vary (strongly) continuously with x_1, it suffices to establish this equality for a dense set of x_1's. Note that the set $D = \{a_0 + \sum_{i=1}^n b_i e_1 c_i : a_0, b_i, c_i \in M, n \in \mathbb{N}\}$ is a self-adjoint subalgebra of M_1 which contains $M \cup \{e_1\}$ and is consequently strongly dense in M_1. Finally it is easily verified that if $x_1 \in D$, then the desired equality is indeed valid.

(ii) Again, we may assume that $x_1 \in D$, as above, and the desired equality is easily verified. □

REMARK 4.3.2 *(a) The proof shows that the preceding lemma is also valid when $N \subseteq M$ is an inclusion of finite-dimensional C^*-algebras, provided the trace we work with is a Markov trace.*

(b) It follows from Lemma 4.3.1 (i) that the set of finite sums of elements of the form $a_0 e_1 b_0$, where $a_0, b_0 \in M$, is an ideal in M_1 and must consequently be equal to all of M_1 by Proposition A.3.1. This shows that there is an isomorphism $\phi : M \otimes_N M \to M_1$ such that $\phi(a_0 \otimes_N b_0) = a_0 e_1 b_0$.

PROPOSITION 4.3.3 *Fix $n \geq [M : N]$.*

(a) Suppose $q \in M_n(N)$ is a projection such that $\operatorname{tr}_{M_n(N)} q = \frac{[M:N]}{n}$. Then there exist $\lambda_1, \cdots, \lambda_n \in M$ such that

$$q_{ij} = E_N(\lambda_i \lambda_j^*) \; \forall i, j. \tag{4.3.2}$$

A collection $\{\lambda_1, \cdots, \lambda_n\} \subset M$ will be called a **basis** *for M/N if the matrix $q = ((q_{ij}))$ defined by equation 4.3.2 above is a projection in $M_n(N)$ such that $\operatorname{tr} q = \frac{[M:N]}{n}$.*

(b) Let $\{\lambda_1, \cdots, \lambda_n\} \subset M$ be basis for M/N. Let $x \in M$ be arbitrary. Then,

(i) $x = \sum_{j=1}^n E_N(x\lambda_j^*)\lambda_j$;

(ii) the row-vector $(E_N(x\lambda_1^*), \cdots, E_N(x\lambda_n^*))$ belongs to $M_{1\times n}(N)q$; and if $(x_1, \cdots, x_n) \in M_{1\times n}(N)q$ satisfies $x = \sum_{i=1}^n x_i \lambda_i$, then $x_j = E_N(x\lambda_j^*) \; \forall j$;

(iii) further, $\sum_{i=1}^n \lambda_i^* e \lambda_i = 1$.

Proof:

(a) Let $q = ((q_{ij})) \in M_n(N)$ be a projection such that $\operatorname{tr} q = \frac{[M:N]}{n}$. Consider the projection $E = ((E_{ij})) \in M_n(M_1)$ defined by $E_{ij} = \delta_{ij} e$; notice that q and E are commuting projections, and so $p = qE$ is also a projection in the II_1 factor $M_n(M_1)$. Observe next that

$$\operatorname{tr} p = \frac{1}{n} \sum_{i=1}^n \operatorname{tr}(q_{ii} e) = \tau \operatorname{tr} q = \frac{1}{n} = \operatorname{tr} e_{11},$$

where e_{11} denotes the projection in $M_n(M_1)$ with (1,1) entry equal to the identity 1, and other entries equal to 0.

Hence the projections p and e_{11} are Murray-von Neumann equivalent in $M_n(M_1)$; thus, there exists a partial isometry $v \in M_n(M_1)$ such that $v^* v = e_{11}$ and $vv^* = p$; the condition $v^* v = e_{11}$ clearly implies that v has the form

$$v = \begin{bmatrix} v_1 & 0 & \cdots & 0 \\ v_2 & 0 & \cdots & 0 \\ \vdots & 0 & \ddots & 0 \\ v_n & 0 & \cdots & 0 \end{bmatrix}, \tag{4.3.3}$$

for uniquely determined $v_1, \cdots, v_n \in M_1$.

By definition of the v_i's, we have

$$\sum_{i=1}^n v_i^* v_i = 1 \tag{4.3.4}$$

and

$$v_i v_j^* = q_{ij} e \ \forall i, j \tag{4.3.5}$$

In particular, note that

$$v_i v_i^* = q_{ii} e \leq e,$$

and hence, we must have $v_i = ev_i$. Then, by the preceding Lemma 4.3.1, there exists a unique $\lambda_i \in M$ such that $v_i = e\lambda_i \ \forall i$.

Deduce that

$$q_{ij} e = e\lambda_i \lambda_j^* e = E_N(\lambda_i \lambda_j^*) e,$$

thereby establishing (a).

Before proceeding to (b), we wish to point out that every 'basis for M/N' arises in the manner indicated in the proof above. (*Reason:* Suppose $\{\lambda_1, \cdots, \lambda_n\} \subset M$ is basis for M/N. Define $v_i = e\lambda_i$, and define $v \in M_n(M_1)$ by equation 4.3.3; it is then seen that $vv^* = qE$, so that v is a partial isometry; also, it follows that the matrix v^*v has 0 entries except at the (1,1) place, and that the (1,1) entry must be f, for some projection $f \in M_1$, which satisfies $\mathrm{tr} f = 1$; in other words, $v^*v = e_{11}$, as desired.)

(b) To start with, note that (iii) is an immediate consequence of equation 4.3.4 (and the fact that $v_i = e\lambda_i$).

As for (i), if $x \in M$ is arbitrary, then

$$\begin{aligned} ex &= \sum_{i=1}^n exv_i^* v_i \\ &= \sum_{i=1}^n ex\lambda_i^* e\lambda_i \\ &= e(\sum_{i=1}^n E_N(x\lambda_i^*)\lambda_i), \end{aligned}$$

and deduce from Lemma 4.3.1(i) that $x = \sum_{i=1}^n E_N(x\lambda_i^*)\lambda_i$.

Note next that if $x \in M$ and if $1 \leq j \leq n$, then,

$$\begin{aligned} e(\sum_{i=1}^n E_N(x\lambda_i^*) q_{ij}) &= \sum_{i=1}^n ex\lambda_i^* e q_{ij} \\ &= \sum_{i=1}^n ex\lambda_i^* e E_N(\lambda_i \lambda_j^*) \\ &= \sum_{i=1}^n ex\lambda_i^* e\lambda_i \lambda_j^* e \\ &= ex\lambda_j^* e \\ &= eE_N(x\lambda_j^*), \end{aligned}$$

and it follows (again from Lemma 4.3.1(i)) that

$$(\sum_{i=1}^n E_N(x\lambda_i^*) q_{ij}) = E_N(x\lambda_j^*);$$

in other words, the row-vector $(E_N(x\lambda_1^*), \cdots, E_N(x\lambda_n^*))$ indeed belongs to $M_{1\times n}(N)q$.

Conversely, if $(x_1, \cdots, x_n) \in M_{1\times n}(N)q$, and if $x = \sum_{i=1}^n x_i\lambda_i$, then observe that

$$\begin{aligned} E_N(x\lambda_j^*) &= \sum_{i=1}^n E_N(x_i\lambda_i\lambda_j^*) \\ &= \sum_{i=1}^n x_i E_N(\lambda_i\lambda_j^*) \\ &= \sum_{i=1}^n x_i q_{ij} \\ &= x_j, \end{aligned}$$

thereby establishing (ii) and hence the proposition. □

A collection $\{\lambda_1, \cdots, \lambda_n\} \subset M$ which satisfies the condition (i) of the above proposition is, strictly speaking, a 'basis' for M, viewed as a left N-module, and should probably be called a 'left-basis' (as against a right-basis, which is what the set of adjoints of a 'left-basis' would be); but we shall stick to this terminology.

The above proposition is a mild extension of the original construction of a basis (in [PP]). (They only discuss the case when the projection q of the proposition has zero entries on the off-diagonal entries.) The reason for our mild extension lies in Lemma 4.3.4(i) and the consequent Remark 4.3.5. We omit the proof of the lemma, which is a routine verification.

LEMMA 4.3.4 *(i) If $N \subseteq M \subseteq P$ is a tower of II_1 factors, with $[P : N] < \infty$, and if $\{\lambda_i : 1 \leq i \leq n\}$ (resp., $\{\eta_j : 1 \leq j \leq m\}$) is a basis for M/N (resp., P/M), then $\{\lambda_i\eta_j : 1 \leq i \leq n, 1 \leq j \leq m\}$ is a basis for P/N.*

(ii) If $\{\lambda_1, \cdots, \lambda_n\}$ is a basis for M/N, then $\{\tau^{-\frac{1}{2}}e_1\lambda_j : 1 \leq j \leq n\}$ is a basis for M_1/M; hence $\{\tau^{-\frac{1}{2}}\lambda_i e_1\lambda_j : 1 \leq i, j \leq n\}$ is a basis for M_1/N. □

REMARK 4.3.5 *Notice, from Lemma 4.3.4(ii), that any element of M_1 is expressible in the form $\sum_{i,j=1}^n a_{ij}\lambda_i e_1\lambda_j$, with the a_{ij}'s coming from N, which is a slightly stronger statement than the fact – see Remark 4.3.2(b) – that elements of the form ae_1b, with $a, b \in M$, linearly span M_1.*

Also, an easy induction argument shows that if $\{\lambda_i : i \in I\}$ is a basis for M/N, and if we define, for $\mathbf{i} = (i_1, \cdots, i_k) \in I^k, k \geq 1$,

$$\lambda_{\mathbf{i}}^{(k)} = \tau^{-\frac{k(k-1)}{4}} \lambda_{i_1} e_1 \lambda_{i_2} e_2 e_1 \lambda_{i_3} \cdots \lambda_{i_{k-1}} e_{k-1} \cdots e_1 \lambda_{i_k},$$

then $\{\lambda_{\mathbf{i}}^{(k)} : \mathbf{i} \in I^k\}$ is a basis for M_{k-1}/N.

Notice next, that if we write $\mathbf{i} \vee \mathbf{j} = (i_1, \cdots, i_k, j_1, \cdots, j_k)$ for $\mathbf{i}, \mathbf{j} \in I^k$, it follows from the commutation relations satisfied by the e_n's – see Proposition 3.3.2 – that

$$\lambda_{\mathbf{i}\vee\mathbf{j}}^{(2k)} = \tau^{-\frac{k^2}{2}} \lambda_{\mathbf{i}}^{(k)} (e_k \cdots e_1)(e_{k+1} \cdots e_2) \cdots (e_{2k-1} \cdots e_k) \lambda_{\mathbf{j}}^{(k)}.$$

This equation, together with the second statement in Lemma 4.3.4, should suggest that we might expect the validity of the next result.

PROPOSITION 4.3.6 *If M_i, e_i are as above, then, for each $m \geq 0, k \geq -1$, the algebra M_{k+2m} is isomorphic to the result of the basic construction applied to the inclusion $M_k \subseteq M_{k+m}$, with a choice of the projection which implements the conditional expectation of M_{k+m} onto M_k being given by*

$$e_{[k,k+m]} = \tau^{-\frac{m(m-1)}{2}}(e_{k+m+1}e_{k+m}\cdots e_{k+2})(e_{k+m+2}\cdots e_{k+3})\cdots(e_{k+2m}\cdots e_{k+m+1}).$$

Proof: This is true basically because of the relations satisfied by the projections e_n. As for the proof, there is no loss of generality in assuming that $k = -1$. Rather than going through the proof of the proposition in its full generality, we shall just present the proof when $k = -1, m = 2$; all the ingredients of the general proof are already present here, and the reader should not have too much trouble writing out the proof in its full generality. In any case, the proof may be found in [PP2].

Thus we have to show that $N \subseteq M_1 \subseteq M_3$ is an instance of the basic construction, with a choice of the projection implementing the conditional expectation of M_1 onto N being given by $f = \tau^{-1}e_2e_1e_3e_2$.

First, the fact that e_1 commutes with e_3 implies that $f^* = f$; and

$$f^2 = \tau^{-2}e_2e_1e_3e_2e_3e_1e_2 = \tau^{-1}e_2e_1e_3e_1e_2 = f,$$

so f is indeed a projection.

Next, for any $x_1 \in M_1$, note that

$$\begin{aligned} fx_1f &= \tau^{-2}e_2e_1e_3e_2x_1e_2e_3e_1e_2 \\ &= \tau^{-2}e_2e_1e_3E_M(x_1)e_2e_3e_1e_2 \\ &= \tau^{-1}e_2e_1E_M(x_1)e_3e_1e_2 \\ &= \tau^{-1}e_2e_1E_M(x_1)e_1e_3e_2 \\ &= \tau^{-1}e_2E_N(x_1)e_1e_3e_2, \end{aligned}$$

and hence, indeed, we have

$$fx_1f = E_N(x_1)f \quad \forall x_1 \in M_1.$$

It follows that there exists a (unique) normal homomorphism π from $P = \langle M_1, e_N^{M_1}\rangle$ (the result of the basic construction for the inclusion $N \subseteq M_1$) onto $P_1 = (M_1 \cup \{f\})''$, such that $\pi(e_N^{M_1}) = f$ and $\pi|_{M_1} = \mathrm{id}_{M_1}$. (The normality, as well as the fact that image is a von Neumann algebra, may be deduced from the second statement in Lemma 4.3.4(ii).) Since P is a II_1 factor, it follows that π is an isomorphism and that P_1 is a II_1 subfactor of M_3 such that $[P_1 : M_1] = [M_1 : N] = [M_3 : M_1]$. This means that $[M_3 : P_1] = 1$, and the proof is complete. □

We conclude this section with a proposition, due to Pimsner and Popa, which shows that the basic construction lends itself to a nice 'duality' result, which, among other things, yields the 'periodicity of order two' in the tower of the basic construction, which was referred to in §4.2.

PROPOSITION 4.3.7 *([PP1]) Let $d = [M : N]$. Then, for each $n \geq -1$, there exists an isomorphism of towers:*

$$(M_d(N) \subseteq M_d(M) \subseteq \cdots \subseteq M_d(M_n)) \cong (M_1 \subseteq M_2 \subseteq \cdots \subseteq M_{n+2}).$$

Proof: We prove the case $n = 0$; the general case is proved similarly, by using Remark 4.3.5.

Fix a basis $\{\lambda_1, \cdots, \lambda_n\}$ for M/N. Let $q = \theta(1)$, where $\theta_{ij}(x) = E_N(\lambda_i x \lambda_j^*)$ $\forall x \in M$, and consider the model $qM_n(N)q$ of $M_d(N)$. Define $\phi : M_d(N) \to M_1$ by

$$\phi((a_{ij})) = \sum_{i,j=1}^{n} \lambda_i^* a_{ij} e_1 \lambda_j.$$

An easy computation shows that ϕ is a unital normal homomorphism, which is necessarily injective. As for surjectivity, if $x_1 \in M_1$, deduce from Proposition 4.3.3(b)(iii) that

$$\begin{aligned} x_1 &= (\sum_i \lambda_i^* e_1 \lambda_i) x_1 (\sum_j \lambda_j^* e_1 \lambda_j) \\ &= \sum_{i,j} \lambda_i^* (e_1 \lambda_i x_1 \lambda_j^* e_1) \lambda_j; \end{aligned}$$

but by Lemma 4.3.1(i), there exists $m_{ij} \in M$ such that $m_{ij} e_1 = \lambda_i x_1 \lambda_j^* e_1$. It is then clear that $x_1 = \phi((E_N(m_{ij})))$, thereby establishing the surjectivity of ϕ.

Now apply the conclusion above, to the basis $\{\tau^{-\frac{1}{2}} e_1 \lambda_i\}$ for M_1/M, to find an isomorphism $\phi_1 : M_d(M) \to M_2$ defined by

$$\phi_1((m_{ij})) = \tau^{-1} \sum_{i,j=1}^{n} \lambda_i^* e_1 m_{ij} e_2 e_1 \lambda_j,$$

which restricts on $M_d(N)$ to the map ϕ defined above, thus proving the proposition. □

4.4 Relative commutants *vs* intertwiners

This section is devoted to establishing the equivalence of the two descriptions of the principal graphs given in §4.2.

We shall consistently use the symbols x_n, y_n, etc., to denote elements of M_n.

PROPOSITION 4.4.1 *(i) For each $n \geq 0$, there exists an isomorphism of squares of finite-dimensional C^*-algebras, as follows:*

$$\begin{pmatrix} N' \cap M_{2n} & \subseteq & N' \cap M_{2n+1} \\ \cup & & \cup \\ M' \cap M_{2n} & \subseteq & M' \cap M_{2n+1} \end{pmatrix} \overset{\phi_n}{\cong} \begin{pmatrix} {}_N\mathcal{L}_M(L^2(M_n)) & \subseteq & {}_N\mathcal{L}_N(L^2(M_n)) \\ \cup & & \cup \\ {}_M\mathcal{L}_M(L^2(M_n)) & \subseteq & {}_M\mathcal{L}_N(L^2(M_n)) \end{pmatrix}.$$

(ii) For each $n \geq 0$, there exists an M_n-M-linear unitary operator

$$u_n : ({}_{M_n}L^2(M_n)_N) \otimes_N ({}_NL^2(M)_M) \cong ({}_{M_n}L^2(M_{n+1})_M).$$

(iii) The composite maps

$$\begin{array}{ccccccc} {}_N\mathcal{L}_N(L^2(M_n)) & \overset{\phi_n^{-1}}{\cong} & N' \cap M_{2n+1} & \subseteq & N' \cap M_{2n+2} & \overset{\phi_{n+1}}{\cong} & {}_N\mathcal{L}_M(L^2(M_{n+1})), \\ {}_M\mathcal{L}_N(L^2(M_n)) & \overset{\phi_n^{-1}}{\cong} & M' \cap M_{2n+1} & \subseteq & M' \cap M_{2n+2} & \overset{\phi_{n+1}}{\cong} & {}_M\mathcal{L}_M(L^2(M_{n+1})) \end{array}$$

are given, in either case, by the formula

$$T \mapsto u_n(T \otimes_N \mathrm{id}_{L^2(M)})u_n^*.$$

Proof: (i) Appeal first to Proposition 4.3.6 to note that $N \subseteq M_n \subseteq M_{2n+1}$ (resp., $M \subseteq M_n \subseteq M_{2n}$) is an instance of the basic construction, and hence $M_{2n+1} \cong J_{M_n}N'J_{M_n}$, so that $(N' \cap M_{2n+1}) \cong (N' \cap J_{M_n}N'J_{M_n}) = {}_N\mathcal{L}_N(L^2(M_n))$. Similarly we can argue that each of the four corners in each square of algebras, displayed in (i) above, is isomorphic to the corresponding corner in the other square. Completing the proof is just a matter of ensuring that the various isomorphisms are compatible. For this, use Lemma 4.3.1(ii) and Proposition 4.3.6 to find that if we define $\phi_n : M_{2n+1} \to \mathcal{L}(L^2(M_n))$ by

$$(\phi_n(x_{2n+1}))(x_n\Omega_{M_n}) = \tau^{-n-1}E_{M_n}(x_{2n+1}x_ne_{[-1,n]})\Omega_{M_n},$$

then $\phi_n(M_{2n+1}) = J_{M_n}N'J_{M_n}$ and $\phi_n|_{M_n} = \mathrm{id}_{M_n}$. A pleasant exercise involving the commutation relations satisfied by the e_n's, coupled with a judicious use of a combination of Lemma 4.3.1(i) and Proposition 4.3.6, as well as such identities as

$$e_{[k,k+m]} = \tau^{-(m-1)}e_{[k+1,k+m]}e_{k+2}e_{k+m+3}\cdots e_{k+m}(e_{k+2m}\cdots e_{k+m+1}),$$

leads to the fact that if $x_{2n} \in M_{2n}$, then

$$(\phi_n(x_{2n}))(x_n\Omega_{M_n}) = \tau^{-n}E_{M_n}(x_{2n}x_ne_{[0,n]})\Omega_{M_n}, \tag{4.4.1}$$

and consequently, as before, $\phi_n(M_{2n}) = J_{M_n}M'J_{M_n}$, and this ϕ_n does all that it is supposed to.

(ii) Recall that if P is any II_1 factor, and if P_0, Q_0 are any two subfactors of finite index in P, and if $\mathcal{H} = L^2(P)$, regarded as a P_0-Q_0-bimodule, then

$\mathcal{H}_0 = P$ and the P_0-valued inner product on $\mathcal{H}$ is given by $\langle x, y\rangle_{P_0} = E_{P_0}(xy^*)$. Define $u_n : M_n \otimes M \to M_{n+1}$ by

$$u_n(x_n \otimes x_0) = \tau^{-\frac{n+1}{2}} x_n e_{n+1} \cdots e_1 x_0.$$

It is a consequence of Remark 4.3.5 that u_n is surjective. An easy computation shows that

$$\begin{aligned} &\langle u_n(x_n \otimes x_0), u_n(y_n \otimes y_0)\rangle_{M_n} \\ &= \tau^{-(n+1)} E_{M_n}(x_n e_{n+1} \cdots e_1 x_0 y_0^* e_1 \cdots e_{n+1} y_n^*) \\ &= x_n E_N(x_0 y_0^*) y_n^* \\ &= \langle x_n \cdot \langle x_0, y_0\rangle_N, y_n\rangle_{M_n}; \end{aligned}$$

it follows from Proposition 4.1.2 that u_n extends to an M_n-M-linear unitary operator of $({}_{M_n}L^2(M_n)_N) \otimes_N ({}_{M_n}L^2(M)_M)$ onto ${}_{M_n}L^2(M_{n+1})_M$, as desired.

(iii) Note, in general, that if $\mathcal{H}$ (resp., $\mathcal{K}$) is an N-P- (resp., P-Q-) bimodule, where N, P, Q are II_1 factors, and if $T \in {}_N\mathcal{L}_P(\mathcal{H}), S \in {}_P\mathcal{L}_Q(\mathcal{K})$, then there is a unique element $T \otimes_P S$ of ${}_N\mathcal{L}_Q(\mathcal{H} \otimes_P \mathcal{K})$ satisfying $(T \otimes_P S)(\xi \otimes_P \eta) = T\xi \otimes_P S\eta, \ \ \forall \xi \in \mathcal{H}_0, \eta \in \mathcal{K}_0$.

Hence, in order to establish the assertion for either of the composite maps, we find, when all the definitions have been unravelled, that it suffices to verify that for arbitrary $x_j \in M_j, j \in \{0, n, 2n+1\}$, we have

$$\phi_{n+1}(x_{2n+1}) \tau^{\frac{n+1}{2}} u_n(x_n \otimes_N x_0) = \tau^{\frac{n+1}{2}} u_n(\phi_n(x_{2n+1}) x_n \otimes_N x_0). \tag{4.4.2}$$

The definitions imply that the right side of equation (4.4.2) is equal to

$$\tau^{-n-1} E_{M_n}(x_{2n+1} x_n e_{[-1,n]}) e_{n+1} \cdots e_1 x_0;$$

on the other hand, it follows from equation (4.4.1) that the left side of equation (4.4.2) is equal to

$$\begin{aligned} &\tau^{-(n+1)} E_{M_{n+1}}(x_{2n+1} x_n e_{n+1} \cdots e_1 x_0 e_{[0,n+1]}) \\ &= \tau^{-(n+1)} E_{M_{n+1}}(x_{2n+1} x_n e_{n+1} \cdots e_1 e_{[0,n+1]}) x_0. \end{aligned}$$

Hence it suffices to prove that, for arbitrary $y_{2n+1} \in M_{2n+1}$, we have

$$E_{M_{n+1}}(y_{2n+1} e_{n+1} \cdots e_1 e_{[0,n+1]}) = E_{M_n}(y_{2n+1} e_{[-1,n]}) e_{n+1} \cdots e_1.$$

An appeal to Proposition 4.3.6 and Lemma 4.3.1 shows that this amounts to showing that

$$E_{M_{n+1}}(y_{2n+1} e_{n+1} \cdots e_1 e_{[0,n+1]}) e_{[0,n+1]} = E_{M_n}(y_{2n+1} e_{[-1,n]}) e_{n+1} \cdots e_1 e_{[0,n+1]},$$

i.e., that

$$\tau^{-n-1} y_{2n+1} e_{n+1} \cdots e_1 e_{[0,n+1]} = E_{M_n}(y_{2n+1} e_{[-1,n]}) e_{n+1} \cdots e_1 e_{[0,n+1]}.$$

On the other hand, it follows immediately from the formula given in Proposition 4.3.6 that

$$e_{n+1} \cdots e_1 e_{[0,n+1]} = e_{[-1,n]} e_{2n+2} \cdots e_{n+2}.$$

Hence it follows, again from Lemma 4.3.1, that

$$E_{M_n}(y_{2n+1} e_{[-1,n]}) e_{n+1} \cdots e_1 e_{[0,n+1]} = \tau^{-n-1} y_{2n+1} e_{[-1,n]} e_{2n+2} \cdots e_{n+2},$$

and a final appeal to Lemma 4.3.1 completes the proof. □

Before proceeding further, notice that if $\mathcal{H}$ is a bifinite P-Q-bimodule, where P, Q are arbitrary II_1 factors, then ${}_P\mathcal{L}_Q(\mathcal{H})$ is a finite-dimensional C^*-algebra. (Reason: Assume $\mathcal{H} = \mathcal{H}_\theta$, for some co-finite morphism $\theta : Q \to M_d(P)$, where $d = \dim_{P-} \mathcal{H}$; then ${}_P\mathcal{L}_Q(\mathcal{H}_\theta) \cong (M_d(P) \cap \theta(Q)'$, which is finite-dimensional in view of the co-finiteness of θ.)

Suppose, for notational convenience, that $A = {}_P\mathcal{L}_Q(\mathcal{H})$. It must be clear that there is a bijective correspondence between $\mathcal{P}(A)$ and the collection of P-Q-sub-bimodules of $\mathcal{H}$ (given by $p \mapsto \operatorname{ran} p$). It follows that, under the above correspondence, the (canonical) decomposition of the identity as a sum of minimal central projections in A corresponds to the (canonical) decomposition of $\mathcal{H}$ into its *isotypical* components; and that, the further (non-canonical) decomposition of a minimal central projection as a sum of minimal projections of A corresponds to the (non-canonical) decomposition of the corresponding isotypical submodule of $\mathcal{H}$ as a direct sum of mutually equivalent irreducible submodules. It follows that if $\mathcal{H}$ is a bifinite P-Q-bimodule, then $\mathcal{H}$ is expressible as the direct sum of finitely many irreducible P-Q-bimodules, and that if $\mathcal{K}$ is an irreducible bifinite P-Q-bimodule, then there is a well-defined multiplicity with which '$\mathcal{K}$ occurs in $\mathcal{H}$'.

If we apply the foregoing remarks to the case when $\mathcal{H} = L^2(M_n), P = Q = N$, we find from Proposition 4.4.1 that the minimal central projections of $(N' \cap M_{2n+1})$ are in bijective correspondence with isomorphism classes of irreducible N-N-sub-bimodules of $L^2(M_n)$. Similarly, the minimal central projections of $(N' \cap M_{2n})$ are in bijective correspondence with isomorphism classes of irreducible N-M-sub-bimodules of $L^2(M_n)$. Further, Proposition 4.4.1(i) implies that in the Bratteli diagram for the inclusion $(N' \cap M_{2n}) \subseteq (N' \cap M_{2n+1})$, the vertex indexed by an irreducible N-M-sub-bimodule X of $L^2(M_n)$, is connected to the irreducible N-N-sub-bimodule Y of $L^2(M_n)$, by k bonds, where k is the multiplicity with which X, when viewed as an N-N-bimodule, occurs in Y. Similarly, the assertion (iii) of Proposition 4.4.1 can be re-interpreted as a version of the Frobenius reciprocity theorem, thus: the inclusion matrix for the inclusion $(N' \cap M_{2n+1}) \subseteq (N' \cap M_{2n+2})$ corresponds to 'induction', in the same way that the inclusion matrix for the inclusion $(N' \cap M_{2n}) \subseteq (N' \cap M_{2n+1})$ corresponds to 'restriction'.

Similar remarks apply also for the relative commutants of M in the members of the tower of the basic construction, and we have completed the justification for the alternative description of the principal and dual graphs, in terms of bimodules, which is given in §4.2.

Chapter 5

Commuting squares

5.1 The Pimsner–Popa inequality

The 'inequality' referred to in the title of this section is actually not an inequality, but a 'minimax' characterisation of the index of a subfactor which can be taken as an alternative definition of the index. (This definition has the advantage of making sense for any inclusion of von Neumann algebras for which there is a conditional expectation of the bigger algebra onto the smaller.)

Before getting to the inequality, we pause to record a lemma – see [Jon1] – which shows that the basic construction is 'generic'.

LEMMA 5.1.1 *If $M_0 \subset M_1$ is an inclusion of II_1 factors with $[M_1 : M_0] < \infty$, then there exists a subfactor M_{-1} such that $M_1 = \langle M_0, e_{M_{-1}} \rangle$.*

Proof: The first step in the proof is to realise $L^2(M_0)$ as an M_1-module in such a way that the action of M_0 agrees with the standard action. To do this, start with a projection $p_1 \in M_1$ satisfying $\operatorname{tr} p_1 = [M_1 : M_0]^{-1}$, observe that $L^2(M_1)p_1$ is an M_0-module with $\dim_{M_0}(L^2(M_1)p_1) = 1$, and use an M_0-linear unitary operator u from $L^2(M_0)$ onto $L^2(M_1)p_1$ (noting that $L^2(M_1)p_1$ is actually an M_1-submodule of $L^2(M_1)$) to transfer the action of M_1 to $L^2(M_0)$, as desired.

So we may assume that $M_1 \subseteq \mathcal{L}(L^2(M_0))$. Now define $M_{-1} = JM_1'J$, where of course J denotes the modular conjugation operator on $L^2(M_0)$. Note then that

$$[M_0 : M_{-1}] = [M'_{-1} : M'_0] = [JM'_{-1}J : JM'_0J] = [M_1 : M_0].$$

If e_1 denotes the orthogonal projection of $L^2(M_0)$ onto $L^2(M_{-1})$, then $e_1 \in M'_{-1}$, and so $e_1 = Je_1J \in M_1$. It follows that if $P = \langle M_0, e_1 \rangle$, then $P \subseteq M_1$ and $[M_1 : P] = 1$; an appeal to Corollary 3.1.4 completes the proof. □

We first state a weaker version of the inequality (which we prove), and later state the stronger version (which we do not prove).

PROPOSITION 5.1.2 *(The Pimsner–Popa inequality) If $N \subseteq M$ is an inclusion of II_1 factors with finite index, if E_N denotes the unique trace-preserving conditional expectation onto N, and if we write $\tau = [M:N]^{-1}$, then*

$$\tau = \sup\{\lambda > 0 : E_N(x) \geq \lambda x \quad \forall x \in M_+\}. \tag{5.1.1}$$

Proof: In the course of the proof, we shall need the following simple fact, which is an immediate consequence of equation (3.1.1), for instance:

The trace-preserving conditional expectation of a finite von Neumann algebra onto a von Neumann subalgebra is *completely positive* – meaning that if $P \subseteq N$ is an inclusion of finite von Neumann algebras, and if $((a_{ij})) \in (M_n(N))_+, n \in \mathbb{N}$, then $((E_P(a_{ij}))) \in (M_n(P))_+$.

To prove the proposition, begin by appealing to Lemma 5.1.1, and assume that $M = \langle N, e_P \rangle$, for some subfactor P of N. Let $x \in M_+$; write $x = zz^*$ for some z in M. By Remark 4.3.2(b), we may assume that $z = \sum_{i=1}^n x_i e_P y_i$ for some $x_i, y_i \in N, n \in \mathbb{N}$. An easy computation now shows that

$$\begin{aligned} zz^* &= \sum_{i,j=1}^n x_i E_P(y_i y_j^*) e_P x_j^*, \\ E_N(zz^*) &= \tau \sum_{i,j=1}^n x_i E_P(y_i y_j^*) x_j^*. \end{aligned}$$

It follows from the complete positivity of E_P (referred to at the start of this proof) that the matrix $y' = ((E_P(y_i y_j^*)))$ is positive. Since $e_P \in P' \cap M$, it follows that

$$\begin{aligned} zz^* &= [x_1 \cdots x_n] y'^{\frac{1}{2}} \begin{bmatrix} e & \cdots & 0 \\ \vdots & \ddots & \vdots \\ 0 & \cdots & e \end{bmatrix} y'^{\frac{1}{2}} \begin{bmatrix} x_1^* \\ \vdots \\ x_n^* \end{bmatrix} \\ &\leq [x_1 \cdots x_n] y' \begin{bmatrix} x_1^* \\ \vdots \\ x_n^* \end{bmatrix} \\ &= \tau^{-1} E_N(zz^*). \end{aligned}$$

Thus we see that $E_N x \geq \tau x \quad \forall x \in M_+$. On the other hand, we know that $E_N(e_P) = \tau 1$. The previous two sentences complete the proof of the proposition. □

Now we state, without proof, the stronger version (see [PP1]), which states that the previous proposition is valid without the assumption of finite index.

THEOREM 5.1.3 *If $N \subseteq M$ is an inclusion of II_1 factors, and if E_N denotes the unique trace-preserving conditional expectation onto N, then*

$$[M:N] = \left(\sup\{\lambda > 0 : E_N(x) \geq \lambda x \quad \forall x \in M_+\}\right)^{-1}. \tag{5.1.2}$$

We single out the amount by which Theorem 5.1.3 is stronger than Proposition 5.1.2, as a separate corollary.

COROLLARY 5.1.4 *If $N \subseteq M$ is an inclusion of II_1 factors, then $[M : N] < \infty$ if and only if there exists a positive constant λ such that $E_N x \geq \lambda x \ \ \forall x \in M_+$.*

The next proposition is an important first step in the computation, by approximation, of the index of a subfactor.

PROPOSITION 5.1.5 *Suppose $N \subseteq M$ is an inclusion of II_1 factors. Suppose that the subfactor and factor can be simultaneously approximated by a grid of von Neumann subalgebras of M in the following sense (where 'convergence' means that the 'limit' algebra is the weak closure of the union of the increasing sequence of subalgebras that form the members of the sequence of 'approximants'):*

$$\begin{array}{ccccccccc} B_0 & \subseteq & B_1 & \subseteq \cdots \subseteq & B_n & \subseteq & \cdots & \to & M \\ \cup & & \cup & & \cup & & & & \cup \\ A_0 & \subseteq & A_1 & \subseteq \cdots \subseteq & A_n & \subseteq & \cdots & \to & N. \end{array}$$

Define

$$\lambda_n = \lambda(B_n, A_n) = \sup\{\lambda > 0 : E_{A_n} x \geq \lambda x \ \ \forall x \in B_{n+}\};$$

then

$$[M : N] \leq (\limsup_n \lambda_n)^{-1}.$$

Proof: Consider the simple fact that if $\{\mathcal{H}_n\}_{n=1}^{\infty}$ is an increasing sequence of closed subspaces of a Hilbert space $\mathcal{H}$, whose union is dense in $\mathcal{H}$, and if p_n denotes the orthogonal projection onto $\mathcal{H}_n$, then $p_n \xi \to \xi \ \ \forall \xi \in \mathcal{H}$. This fact has the following consequences in the context of this proposition:

(i) $E_{A_m} z \to E_N z$ weakly, for all z in M; and

(ii) if $x \in M$ and $x_n = E_{B_n} x$, then $x_n \to x$ strongly.

Fix $x \in M_+$, and let x_n be as in (ii) above. Let $\{r_m\}$ be a sequence in $(0, 1)$ such that $r_m \to 1$ as $m \to \infty$. Now fix an integer n, and note that since $x_n \in B_{m+} \ \forall m \geq n$, the definition of λ_m ensures that $E_{A_m}(x_n) \geq r_m \lambda_m x_n \ \forall m \geq n$. Now use (i) and (ii) above, and conclude, by first letting $m \to \infty$ and then letting $n \to \infty$, that $E_N x \geq (\limsup_n \lambda_n) x$; this completes the proof, in view of Theorem 5.1.3. □

EXAMPLE 5.1.6 *Let* $B_0 \overset{\Lambda}{\subseteq} B_1$ *be a connected inclusion of finite-dimensional* C^**-algebras, and let* $B_0 \subseteq B_1 \subseteq B_2 \subseteq \cdots \subseteq B_n \subseteq \cdots$ *denote the tower of the basic construction with respect to the Markov trace. Then, as has already been observed, there is a unique trace on* $\bigcup_n B_n$ *and the von Neumann algebra completion of this union is the hyperfinite* II_1 *factor* R.

Set $B_{-1} = \mathbb{C}$, *and pick arbitrary unitary elements* $u_n \in B'_{n-1} \cap B_{n+1}$; *it follows exactly as in Example 3.1.6 that there exists a unique normal endomorphism* $\alpha : R \to R$ *with the property that* $\alpha|_{B_n} = \mathrm{Ad}_{u_1 u_2 \cdots u_n}|_{B_n}$. *(This construction is fairly general; for instance, if each* B_n *is a factor, then this is the form of the most general endomorphism of* $\bigcup_n B_n$ *which maps* B_n *into* B_{n+1} *for all* n.*)*

Set $A_n = \alpha(B_{n-1}), n \geq 0, M = R, N = \alpha(R)$, *and note that for* $n \geq 1$, *we have*

$$(A_n \subseteq A_{n+1} \subseteq B_{n+1}) = \mathrm{Ad}_{u_1 u_2 \cdots u_n} (B_{n-1} \subseteq B_n \subseteq B_{n+1}),$$

and consequently, $(A_n \subseteq A_{n+1} \subseteq B_{n+1})$ *is an instance of the basic construction; it follows from this and the proof of Proposition 5.1.2 that, in the notation of Proposition 5.1.5, we have, for all* $k \geq 2$,

$$\lambda_k = ||\Lambda||^{-2}.$$

Thus, we find, by Proposition 5.1.5, that no matter what the initial sequence $\{u_n\}$ *of unitary elements is, we always have* $[M : N] \leq ||\Lambda||^2 < \infty$.

It is known – see [Jon6] and [Ake], for instance – from examples that the exact value of the index could be just about any admissible index value in the range $[1, ||\Lambda||^2]$; *it would be very interesting to determine the exact dependence of the index on the initial sequence of unitary elements.*

The possible usefulness of Proposition 5.1.5 becomes enhanced in the light of a result, due to Popa, to the effect that if $N \subset M$ is an inclusion of hyperfinite II_1 factors, then it is always possible to find *finite-dimensional* C^*-subalgebras A_n, B_n of M satisfying the conditions of the above proposition.

Further, Proposition 5.1.5 would be even more useful, if we could sharpen the inequality to an equality. It becomes clear very quickly, however, that it is unreasonable to expect any such equality without further conditions on the approximating grid; for instance if $\{A_n\}$ is any increasing sequence of finite-dimensional C^*-subalgebras of R whose union is weakly dense in R, and if we set $B_n = A_{n+k}$ for some arbitrarily fixed k, then in the notation of Proposition 5.1.5, we would have $N = M = R$, while the numbers λ_n could all be quite large!

This leads us to the very important notion of a 'commuting square of finite von Neumann algebras', which, as we shall soon see, very satisfactorily plays the role of the 'further conditions' that were mentioned above.

DEFINITION 5.1.7 *Suppose* B_1 *is a finite von Neumann algebra, and* tr *is a faithful normal tracial state on* B_1. *Suppose* A_0, A_1 *and* B_0 *are von Neumann* *subalgebras of B_1 such that $A_0 \subseteq A_1 \cap B_0$, and suppose the following clearly equivalent conditions are satisfied:*

(i)

$$\begin{array}{ccc} B_0 & \hookrightarrow & B_1 \\ E_{A_0} \downarrow & & \downarrow E_{A_1} \\ A_0 & \hookrightarrow & A_1 \end{array}$$

is a commutative diagram of maps;

(i)' $E_{A_1}(B_0) = A_0$;

(ii)

$$\begin{array}{ccc} A_1 & \hookrightarrow & B_1 \\ E_{A_0} \downarrow & & \downarrow E_{B_0} \\ A_0 & \hookrightarrow & B_0 \end{array}$$

is a commutative diagram of maps;

(ii)' $E_{B_0}(A_1) = A_0$;

(iii) $E_{A_1} E_{B_0} = E_{A_0}$;

(iii)' $E_{B_0} E_{A_1} = E_{A_0}$.

When the preceding conditions are satisfied, we shall (often suppress specific reference to the trace tr*, and simply) say that*

$$\begin{array}{ccc} B_0 & \subseteq & B_1 \\ \cup & & \cup \\ A_0 & \subseteq & A_1 \end{array}$$

is a commuting square of finite von Neumann algebras.

We shall discuss several examples of such commuting squares in the next section. Before that, however, we conclude this section by making a general remark about the notion of a commuting square, and then fulfil the assertion made earlier about this notion by showing that in the presence of the commuting square condition, we can sharpen Proposition 5.1.5 to the desired equality.

REMARK 5.1.8 *Let us temporarily adopt the convention that if $\mathcal{G}$ is a directed graph without multiple bonds, then by a representation of $\mathcal{G}$ we shall mean an assignment of an algebra A_v to each vertex v in the graph $\mathcal{G}$ and inclusion maps $i_e : A_{v_1} \to A_{v_2}$ whenever e is an edge in $\mathcal{G}$ from v_1 to v_2. It can then be shown that any tree admits an essentially unique representation, while if $\mathcal{G}$ has cycles, there are some 'parameters' or in other words, there is a 'moduli space'. A commuting square may be considered as a point in this moduli space corresponding to the graph given by a square.*

PROPOSITION 5.1.9 *Suppose $M, N, A_n, B_n, \lambda_n$ are as in Proposition 5.1.5, and suppose further that*

$$\begin{array}{ccc} B_n & \subseteq & B_{n+1} \\ \cup & & \cup \\ A_n & \subseteq & A_{n+1} \end{array}$$

is, for each n, a commuting square. Then

$$[M : N] = (\lim_{n\to\infty} \lambda_n)^{-1}.$$

Proof: The point to observe is that the result of stacking up (finitely or countably many) commuting squares is again a commuting square. The necessary formal argument is simple (and uses essentially only the simple fact about increasing sequences of projections in Hilbert space which was mentioned at the start of the proof of Proposition 5.1.5), and may be safely omitted. □

5.2 Examples of commuting squares

In this section, we shall discuss three families of examples of commuting squares. Each of them is of the special form

$$\begin{array}{ccc} uA_1u^* & \subseteq & A_2 \\ \cup & & \cup \\ A_0 & \subseteq & A_1 \end{array} \tag{5.2.1}$$

where (i) $A_0 \subseteq A_1$ is a 'connected' inclusion of finite-dimensional C^*-algebras, (ii) $A_2 = \langle A_1, e_2 \rangle$ is the result of the basic construction applied to $A_0 \subseteq A_1$; and (iii) u is a suitable unitary element in $A_0' \cap A_2$ such that (5.2.1) is a commuting square, with respect to the Markov trace on A_2.

5.2.1 The braid group example

Consider the square (5.2.1), where $u = (q+1)e_2 - 1$, for some scalar q which must necessarily satisfy $|q| = 1$ (in order for u to be unitary). It is an easy exercise to check that for such a u, the square (5.2.1) satisfies the commuting square condition if and only if

$$2 + q + q^{-1} = (\operatorname{tr} e_2)^{-1}.$$

On the other hand, since the condition $|q| = 1$ clearly implies that $|2 + q + q^{-1}| < 4$, this construction can work only when $\tau^{-1} = 4\cos^2 \frac{\pi}{m}$, for some $m \geq 3$.

The reason for calling this the 'braid group example' is clarified in §A.5 of the appendix.

5.2.2 Spin models

The reason that these examples are called *spin models* stems from statistical mechanical considerations – see, for instance [Jon3].

Consider the square (5.2.1), where we now assume that $A_0 = \mathbb{C}$ and $A_1 = \Delta$ is the diagonal subalgebra of $A_2 = M_N(\mathbb{C})$ – see Example 3.3.1(ii). Another easy computation shows that in this case, the square (5.2.1) satisfies the commuting square condition if and only if $|u_{ij}| = \frac{1}{\sqrt{N}} \; \forall i, j$.

A square matrix u, as above – i.e., a unitary matrix, all of whose entries have the same modulus – is called a *complex Hadamard matrix*. The simplest example of such a matrix is given, for any N, by $u_{ij} = \frac{1}{\sqrt{N}} \exp(2\pi\sqrt{-1}ij) = \frac{\omega^{ij}}{\sqrt{N}}$, where ω denotes a primitive N-th root of unity. (This matrix may be called the finite Fourier transform, as it diagonalises the left regular representation of the cyclic group $\mathbb{Z}_N$.) To see that there are many complex Hadamard matrices, observe the following one-parameter family of 4×4 complex Hadamard matrices, indexed by the parameter t ranging over the unit circle $\mathbb{T}$ in $\mathbb{C}$:

$$u_t = \frac{1}{2} \begin{bmatrix} 1 & t & -1 & t \\ t & 1 & t & -1 \\ -1 & t & 1 & t \\ t & -1 & t & 1 \end{bmatrix}.$$

The reason for the adjective 'complex' in the previous paragraph is that there is a fair amount of history related to real orthogonal matrices with constant modulus – these were studied by Hadamard and bear his name. We record here a few facts about such real Hadamard matrices:

(i) Call a positive integer N an Hadamard integer if there exists an $N \times N$ real Hadamard matrix. It is easy to show that if N is an Hadamard integer and if $N > 2$, then $N \equiv 0 (\text{mod}\, 4)$. The open question concerning real Hadamard matrices is whether every multiple of 4 is in fact an Hadamard integer. (This has been verified to be true for some fairly large multiples of 4.)

(ii) It is not hard to show that two squares of the form (5.2.1) that are given by real Hadamard matrices are isomorphic (as squares of algebras) if and only if the corresponding Hadamard matrices are equivalent in the sense that it is possible to obtain one by (pre- and post-) multiplying the other by diagonal orthogonal matrices and permutation matrices. The number h of isomorphism classes of real Hadamard matrices of size N, for small N, is given as follows:

N	2	4	8	12	16
h	1	1	1	1	5

5.2.3 Vertex models

These examples also get their names from statistical mechanical considerations – see [Jon3].

Consider the square (5.2.1), where we now assume, as in Example 3.3.1(i), that $A_0 = \mathbb{C}, A_1 = M_N(\mathbb{C}) \otimes 1, A_2 = M_N(\mathbb{C}) \otimes M_N(\mathbb{C})$. If we make the identification $A_i = M_{N^i}(\mathbb{C})$, and if $u = ((u^{ij}_{kl}))_{1 \leq i,j,k,l \leq N}$, then the square (5.2.1) satisfies the commuting square condition if and only if the matrix u is *biunitary*, meaning that it satisfies the 'two unitarity conditions':

$$\sum_{i,j} u^{ij}_{kl} \overline{u^{ij}_{k'l'}} = \delta_{k,k'} \delta_{l,l'},$$

$$\sum_{i,k} u^{ij}_{kl} \overline{u^{ij'}_{kl'}} = \delta_{j,j'} \delta_{l,l'}.$$

The simplest example of a vertex model is provided by the 'flip' – which was briefly encountered in Example 3.1.6 – for which $u^{ij}_{kl} = \delta_{il}\delta_{kj}$.

We shall consider the last two classes of examples in greater detail in Chapter 6 (where, among other things, we shall see the reason for the above 'biunitarity' condition).

5.3 Basic construction in finite dimensions

Recall that if e is a projection in a von Neumann algebra, then the projection $\bigvee\{ueu^* : u \in M, u \text{ unitary}\}$ is also given by $\bigwedge\{f \in \mathcal{P}(Z(M)) : e \leq f\}$; this projection, which we shall denote by $z_M(e)$, is called the *central support* of the projection e.

LEMMA 5.3.1 *Let $A \subseteq B$ be a unital inclusion of finite-dimensional C^*-algebras. Fix a faithful tracial state ϕ on B, and let J denote the modular conjugation operator on $L^2(B, \phi)$. Let e denote the orthogonal projection of $L^2(B, \phi)$ onto the subspace $L^2(A, \phi|_A)$; let E_A denote the ϕ-preserving conditional expectation of B onto A; and let $B_1 = \langle B, e \rangle$ denote the result of the basic construction.*

(a) $z_{B_1}(e) = 1$.

(b) Suppose C is a finite-dimensional C^*-algebra containing B. (We only consider unital inclusions.) Suppose C contains a projection f satisfying:

(i) $C = \langle B, f \rangle$;

(ii) $fbf = E_A(b)f, \quad \forall b \in B$; and

(iii) $a \mapsto af$ is an injective map of A into C.

Then $BfB = CfC = Cz_C(f)$, and there exists a unique isomorphism $\psi : B_1 \to CfC$ such that $\psi(e) = f$ and $\psi(b) = bz_C(f) \quad \forall b \in B$.

Proof: (a) Suppose $z \in \mathcal{P}(Z(B_1))$ and $e \leq z$. Then also $e = JeJ \leq JzJ \in \mathcal{P}(Z(JB_1J)) = \mathcal{P}(Z(A))$; but then $(1 - JzJ)e = 0$, and since the map $a \mapsto ae$ is an injective map of A, it follows, as desired, that $z = 1$.

(b) The assumptions clearly imply that $C = \{b_0 + \sum_{i=1}^n b_i f b_i' : n \in \mathbb{N}, b_i, b_i' \in B\}$. It follows – from this and (ii) – that $CfC = BfB$.

On the other hand, every two-sided ideal in the finite-dimensional C^*-algebra is of the form Cz, for a uniquely determined central projection $z \in C$. It follows from the definition of the central support that $CfC = Cz_C(f)$.

In order to complete the proof, we need to show that the assignment $\sum_{i=1}^n b_i e b_i' \mapsto \sum_{i=1}^n b_i f b_i'$ yields a well-defined injective mapping from B_1 to CfC. (It will immediately follow that such a map would be the desired isomorphism.) That this assignment is indeed an unambiguously defined bijection to CfC from B_1 is an immediate consequence of (two applications, once to C, and once to B_1, of) the following:

Assertion: Let $n \in \mathbb{N}, b_i, b_i' \in B, 1 \leq i \leq n$. Then

$$\sum_{i=1}^n b_i f b_i' = 0 \Leftrightarrow \sum_{i=1}^n E_A(bb_i)E_A(b_i'b') = 0 \quad \forall b, b' \in B. \tag{5.3.1}$$

Proof of assertion: Suppose $\sum_{i=1}^n b_i f b_i' = 0$. Pre-multiply by fb and post-multiply by $b'f$, and appeal to the hypotheses (ii) and (iii), to conclude that indeed, $\sum_{i=1}^n E_A(bb_i)E_A(b_i'b') = 0 \ \ \forall b, b' \in B$. Conversely, if the latter condition is satisfied, pre-multiply by xf and post-multiply by fx', to deduce that $xfb(\sum_{i=1}^n b_i f b_i')b'fx' = 0$, for arbitrary $x, x' \in B$. Since the finite-dimensional C^*-algebra $CfC = BfB$ has a unit, this implies that indeed $\sum_{i=1}^n b_i f b_i' = 0$, and the proof of the assertion, and consequently of the lemma, is complete. □

COROLLARY 5.3.2 *Let A, B, B_1, C, e, f be as in Lemma 5.3.1(b). Then the following conditions are equivalent:*

(i) there exists an isomorphism $\phi : B_1 \to C$ such that $\phi|_B = \mathrm{id}_B$ and $\phi(e) = f$;

(ii) $z_C(f) = 1$.

Proof: Obvious. □

We come now to a very important relation between commuting squares and the basic construction.

LEMMA 5.3.3 *Suppose*

$$\begin{array}{ccc} B_0 & \overset{H}{\subset} & B_1 \\ {}_K\cup & & \cup_L \\ A_0 & \underset{}{\overset{G}{\subset}} & A_1 \end{array} \tag{5.3.2}$$

is a commuting square of finite-dimensional C^*-algebras with respect to a trace on B_1 which is a Markov trace for the inclusion $B_0 \subset B_1$, with inclusion matrices as indicated.

Let $B_2 = \langle B_1, e\rangle$ denote the result of the basic construction for the inclusion $B_0 \subset B_1$, where e denotes the projection in B_2 which implements the conditional expectation of B_1 onto B_0. Define $A_2 = \langle A_1, e\rangle \subset B_2$. Then the following is also a commuting square (with respect to the unique trace on B_2 which extends the given trace on B_1 and is a Markov trace for the inclusion $B_1 \subset B_2$), with the inclusion matrices as indicated:

$$\begin{array}{ccc} B_1 & \overset{H'}{\subset} & B_2 \\ {\scriptstyle L}\cup & & \cup{\scriptstyle K_1} \\ A_1 & \overset{G_1}{\subset} & A_2, \end{array} \tag{5.3.3}$$

where G_1 (resp., K_1) is a matrix with block-form $[G' \ \Gamma]$ (resp., $\begin{bmatrix} K \\ \Lambda \end{bmatrix}$), where Γ (resp., Λ) is the submatrix of G_1 (resp., K_1) determined by the columns (resp., rows) indexed by minimal central projections of A_2 which are orthogonal to e.

Proof: Begin by observing that, in view of the assumed commuting square condition, the tower $A_0 \subseteq A_1 \subseteq A_2 = \langle A_1, e\rangle$ satisfies the conditions of Lemma 5.3.1(b); in particular, any element of A_2 is of the form $x = a + \sum_{i=1}^n b_i e c_i$, where $a, b_i, c_i \in A_1$; and $E_{B_1} x = a + \sum_{i=1}^n \tau b_i c_i \in A_1$, where $\tau = ||H||^{-2}$, and consequently (5.3.3) is also a commuting square.

Let $p_1^{(l)}, \cdots, p_{m_l}^{(l)}$ (resp., $q_1^{(l)}, \cdots, q_{n_l}^{(l)}$) be an ordering of the minimal central projections of A_l (resp., B_l) for $l = 0, 1$, with respect to which the inclusion matrices are G, H, K, L as in (5.3.2). If we define $q_k^{(2)} = J_{B_1} q_k^{(0)} J_{B_1}, 1 \leq k \leq n_0$, it follows from Lemma 3.2.2(a)(iv) that $q_1^{(2)}, \cdots, q_{n_0}^{(2)}$ is an ordering of the minimal central projections of B_2, and hence the inclusion matrix for $B_1 \subseteq B_2$, computed with respect to the $q_j^{(1)}$'s and the $q_k^{(2)}$'s, is H'.

On the other hand, it follows from Lemma 5.3.1(b) that, if we write $z = z_{A_2}(e)$, then there exists an isomorphism $\phi : \langle A_1, e_{A_0}^{A_1}\rangle \to A_2 z$ such that $\phi(e_{A_0}^{A_1}) = e$ and $\phi(a_1) = a_1 z \ \ \forall a_1 \in A_1$. Define $p_i^{(2)} = \phi(J_{A_1} p_i^{(0)} J_{A_1}), 1 \leq i \leq m_0$, and let $p_{m_0+1}^{(2)}, \cdots, p_{m_2}^{(2)}$ be a listing of the minimal central projections of A_2 which are orthogonal to z (or equivalently, to e). It follows, again from Lemma 3.2.2, that $p_1^{(2)}, \cdots, p_{m_2}^{(2)}$ is a listing of the minimal central projections of A_2 such that the inclusion matrix for $A_1 \subseteq A_2$, computed with respect to the $p_i^{(1)}$'s and the $p_k^{(2)}$'s, is $[G' \ \Gamma]$, as asserted.

Finally, fix $1 \leq i \leq m_0, 1 \leq k \leq n_0$, and fix a minimal projection $f \in A_0 p_i^{(0)}$. By definition of the inclusion matrix, there exists a decomposition $f q_k^{(0)} = \sum_{s=1}^{K_{ik}} r_s$ of $f q_k^{(0)}$ as a sum of (mutually orthogonal) minimal projections in $B_0 q_k^{(0)}$. Notice that $\phi(f e_{A_0}^{A_1}) = \phi(f)\phi(e_{A_0}^{A_1}) = fz \cdot e = fe$. Hence we may deduce from Lemma 3.2.2(b)(ii) that fe is a minimal projection in $A_2p_i^{(2)}$, and further,

$$\begin{aligned} feq_k^{(2)} &= fq_k^{(2)}e \\ &= fq_k^{(0)}e \qquad \text{(by Lemma 3.2.2(a)(i))} \\ &= \sum_{s=1}^{K_{ik}} r_s e; \end{aligned}$$

but, again by Lemma 3.2.2(b)(ii), each r_se is a minimal projection in $B_2q_k^{(2)}$. This completes the proof that the inclusion matrix for $A_2 \subseteq B_2$ is indeed as asserted. □

COROLLARY 5.3.4 *Let*

$$\begin{array}{ccc} B_0 & \overset{H}{\subset} & B_1 \\ {}_K\cup & & \cup_L \\ A_0 & \overset{G}{\subset} & A_1 \end{array} \tag{5.3.4}$$

be as in Lemma 5.3.3.

(a) Then $G'K \leq LH'$ in the entry-wise sense.

(b) Assume that each of the four inclusions in (5.3.4) is connected. Then the following conditions are equivalent:

(i) $G'K = LH'$;

(ii) if A_2, B_2 are as in Lemma 5.3.3, then $z_{A_2}(e) = 1$;

(iii) $\bigvee\{ueu^* : u \in A_1, u$ unitary$\} = 1$, where the 'supremum' is computed in A_2 or equivalently in $\mathcal{L}(L^2(B_1))$;

(iv) B_1 is linearly spanned by $A_1B_0 = \{a_1b_0 : a_1 \in A_1, b_0 \in B_0\}$;

(i)' $K'G = HL'$;

(ii)' if $C_1 = \langle B_1, f\rangle$ is the result of the basic construction for the inclusion $A_1 \subseteq B_1$ (with respect to the Markov trace for this inclusion), and if $C_0 = \langle B_0, f\rangle$, then $z_{C_0}(f) = 1$;

(iii)' if C_0 and f are as in (ii)' above, then $\bigvee\{ufu^* : u \in B_0, u$ unitary$\} = 1$, where the 'supremum' is computed in C_0 or equivalently in $\mathcal{L}(L^2(B_1))$;

(iv)' B_1 is linearly spanned by $B_0A_1 = \{b_0a_1 : a_1 \in A_1, b_0 \in B_0\}$.

(c) Assume that each of the four inclusions in (5.3.4) is connected, and that the equivalent conditions of (b) are satisfied; then:

(i) $||G|| = ||H||$;

(ii) $||K|| = ||L||$;

(iii) *if* tr *denotes the trace on* B_1 *which is the Markov trace for* $B_0 \subseteq B_1$, *then* tr *is also the Markov trace for* $A_1 \subseteq B_1$, *and* $\mathrm{tr}|_{A_1}$ *(resp.,* $\mathrm{tr}|_{B_0}$*) is the Markov trace for* $A_0 \subseteq A_1$ *(resp.,* $A_0 \subseteq B_0$*.)*

Proof: (a) By multiplicativity of inclusion matrices, we find from Lemma 5.3.3 that ($KH = GL$ and) $LH' = G_1K_1 = G'K + \Gamma\Lambda$, thus establishing (a).

(b) By the above reasoning, if (ii) is satisfied, it follows from Lemma 5.3.3 that Γ and Λ are absent, whence we have (i).

On the other hand, suppose (ii) is violated. In the notation of the proof of Lemma 5.3.3, this means there exists $i, m_0 < i \leq m_2$, such that $p_i^{(2)} \neq 0$. Since the inclusion $A_1 \subseteq A_2$ is unital, there must exist $j \leq m_1$ such that $\Gamma_{ji} \neq 0$. Also, by the definition of an inclusion, there must exist $k, 1 \leq k \leq n_2$, such that $\Lambda_{ik} \neq 0$; it follows that $(\Gamma\Lambda)_{jk} \neq 0$, and hence $(G'K)_{jk} < (LH')_{jk}$, thus establishing that (i) $\Leftrightarrow$ (ii).

(In the following, we shall write Ω for the cyclic trace vector in $L^2(B_1)$.) Notice that the condition (ii) amounts to the requirement that $\bigvee\{ueu^* : u \in A_2, u \text{ unitary}\} = 1$, which amounts to requiring (since the range of ueu^* is just $uB_0\Omega$) that $B_1\Omega$ is spanned by $\bigcup\{uB_0\Omega : u \in A_2, u \text{ unitary}\}$, which is equivalent to the requirement that $B_1\Omega$ is spanned by $\{xb_0\Omega : x \in A_2, b_0 \in B_0\}$. Since A_2 is spanned by A_1 together with elements of the form $aea_1, a, a_1 \in A_1$, and since the range of e is just $B_0\Omega$, it follows that (ii) is equivalent to the requirement that $B_1\Omega$ is spanned by $A_1B_0\Omega$ (which, in view of Ω being a separating vector for B_1, is the same as (iv)) which is again seen to be equivalent (by reversing the earlier reasoning) to (iii).

By taking adjoints, it is clear that the conditions (iv) and (iv)$'$ are equivalent, while the equivalence of (i)$'$–(iv)$'$ follows from the equivalence of (i)–(iv).

(c) The assumption that (b)(ii) is satisfied implies – by Proposition 3.2.3 – that $\mathrm{tr}|_{A_1}$ is the Markov trace for $A_0 \subseteq A_1$ (since $E_{B_1}e = ||H||^{-2}$) and that $||G|| = ||H||$.

As for (iii), let us write t^{B_1} for the trace vector corresponding to the trace tr. Let us also write t^C for the trace vector corresponding to $\mathrm{tr}|_C$, where $C \in \{A_0, A_1, B_0\}$. The already established fact that $\mathrm{tr}|_{A_1}$ is the Markov trace for $A_0 \subseteq A_1$ implies that

$$G't^{A_0} = ||G||^2 t^{A_1}. \tag{5.3.5}$$

Hence

$$H'HL't^{A_1} = H'K'Gt^{A_1} = H'K't^{A_0} = L'G't^{A_0} = ||G||^2L't^{A_1}.$$

Thus, $L't^{A_1}$ is a, and hence the, Perron–Frobenius eigenvector for $H'H$; i.e., there exists a positive constant ρ such that $L't^{A_1} = \rho t^{B_1}$; but then, also $LL't^{A_1} = \rho Lt^{B_1} = \rho t^{A_1}$; and hence $\rho = ||L||^2$, and indeed tr is the Markov trace for $A_1 \subseteq B_1$. □

REMARK 5.3.5 *The proof shows that the conditions (ii),(iii),(iv),(ii)$'$,(iii)$'$, and (iv)$'$ in Corollary 5.3.4(b) are equivalent even without requiring either* *that the trace in question is the Markov trace, or that all the inclusions are connected. In fact, these conditions are equivalent under just the assumption that we have a commuting square of finite von Neumann algebras, provided that the expression 'linearly spanned' in conditions (iv) and (iv)' is replaced by 'contained in the closed subspace of $L^2(B_1)$ generated by'. Such a general commuting square of finite von Neumann algebras has also been called a* non-degenerate *commuting square by Popa.*

DEFINITION 5.3.6 *The commuting square (5.3.4) is called a symmetric commuting square if the equivalent conditions in Corollary 5.3.4(b) are satisfied.*

5.4 Path algebras

Suppose

$$A_0 \subseteq A_1 \subseteq A_2 \subseteq \cdots \subseteq A_n \subseteq \cdots \tag{5.4.1}$$

is a tower of finite-dimensional C^*-algebras. We know that the 'isomorphism class' of this tower is completely determined by the dimension vector of A_0 and the Bratteli diagrams of the successive inclusions. The path-algebra formalism gives a model of such a tower (for any prescribed data as above), which is quite useful in some computations. Before we get to discussing this formalism, we set up some notation.

If A is a finite-dimensional C^*-algebra, we shall write $\pi(A)$ for the set of minimal central projections of A. Thus, if $A \subseteq B$ is a unital inclusion of finite-dimensional C^*-algebras, then the inclusion matrix is the $\pi(A) \times \pi(B)$ matrix, with (p,q)-th entry given by $[\dim_{\mathbb{C}}(pqA'pq \cap pqBpq)]^{\frac{1}{2}}$. We shall think of an edge α in the Bratteli diagram for the inclusion $A \subseteq B$, which joins $p \in \pi(A)$ and $q \in \pi(B)$, say, as being oriented so that it starts from p and finishes at q, and we shall write $p = s(\alpha), q = f(\alpha)$.

Suppose now that the A_k's are as in (5.4.1), where we assume that all the inclusions are unital. For convenience, we set $A_{-1} = \mathbb{C}1 \subseteq A_0$; write $\pi(A_{-1}) = \{*\}$. For $k \geq 0$, let Ω_k denote the set of edges in the Bratteli diagram for the inclusion $A_{k-1} \subseteq A_k$, oriented as discussed in the last paragraph.

For $-1 \leq l < k < \infty$, define

$$\begin{aligned}
\Omega_{[l,k]} &= \{\alpha = (\alpha_{l+1}, \alpha_{l+2}, \cdots, \alpha_k) : \alpha_i \in \Omega_i, f(\alpha_i) = s(\alpha_{i+1})\ \ \forall i\}; \\
\Omega_{[l,\infty]} &= \{\alpha = (\alpha_{l+1}, \alpha_{l+2}, \cdots, \alpha_k, \cdots) : \alpha_i \in \Omega_i, f(\alpha_i) = s(\alpha_{i+1})\ \ \forall i\}; \\
\Omega_{[k,k]} &= \pi(A_k).
\end{aligned}$$

For $-1 \leq l \leq k \leq \infty$, let $\mathcal{H}_{[l,k]}$ denote a Hilbert space with an orthonormal basis indexed by $\Omega_{[l,k]}$; we shall identify an operator x on $\mathcal{H}_{[l,k]}$ with its matrix $((x(\alpha,\beta)))$ (with respect to this basis).

If $-1 \leq l \leq l_1 < k_1 \leq k < \infty$, and if $\alpha = (\alpha_{l+1}, \alpha_{l+2}, \cdots, \alpha_k) \in \Omega_{[l,k]}$, write $s(\alpha) = s(\alpha_{l+1}), f(\alpha) = f(\alpha_k)$ and $\alpha_{[l_1,k_1]} = (\alpha_{l_1+1}, \alpha_{l_1+2}, \cdots, \alpha_{k_1})$.

We shall use the following obvious simplifying notational conventions:

$$\Omega_{k]} = \Omega_{[-1,k]}, \Omega_{[l} = \Omega_{[l,\infty]}, \Omega = \Omega_{[-1};$$

$$\mathcal{H}_{k]} = \mathcal{H}_{[-1,k]}, \mathcal{H}_{[l} = \mathcal{H}_{[l,\infty]}, \mathcal{H} = \mathcal{H}_{[-1};$$

$$\alpha_{k]} = \alpha_{[-1,k]}, \alpha_{[l} = \alpha_{[l,\infty)}.$$

Finally, for $-1 \leq n < \infty$, define

$$B_n = \{x \in \mathcal{L}(\mathcal{H}) : \exists a \in \mathcal{L}(\mathcal{H}_{n]}) \text{ such that } x(\alpha, \beta) = \delta_{\alpha_{[n}, \beta_{[n}} a(\alpha_{n]}, \beta_{n]}), \ \ \forall \alpha, \beta \in \Omega\}.$$

PROPOSITION 5.4.1 *With the foregoing notation, we have:*

(i) for $-1 \leq n < \infty, B_n$ *is a *-subalgebra of* $\mathcal{L}(\mathcal{H})$ *and*

$$B_0 \subseteq B_1 \subseteq B_2 \subseteq \cdots \subseteq B_n \subseteq \cdots; \tag{5.4.2}$$

(ii) if $-1 \leq k \leq n < \infty$, *then*

$$\begin{aligned} B_k' = \{x \in \mathcal{L}(\mathcal{H}) : \exists a \in \mathcal{L}(\mathcal{H}_{[k}) &\text{ such that} \\ x(\alpha, \beta) = \delta_{\alpha_{k]}, \beta_{k]}} a(\alpha_{[k}, \beta_{[k}) &\ \ \forall \alpha, \beta \in \Omega\}, \\ B_k' \cap B_n = \{x \in \mathcal{L}(\mathcal{H}) : \exists a \in \mathcal{L}(\mathcal{H}_{[k,n]}) &\text{ such that} \\ x(\alpha, \beta) = \delta_{\alpha_{k]}, \beta_{k]}} \delta_{\alpha_{[n}, \beta_{[n}} a(\alpha_{[k,n]}, \beta_{[k,n]}) &\ \ \forall \alpha, \beta \in \Omega\}; \end{aligned}$$

(iii) for fixed $n \geq -1$ *and* $p \in \pi(A_n)$, *define* $\tilde{p} \in B_n$ *by*

$$\tilde{p}(\alpha, \beta) = \delta_{\alpha,\beta} \delta_{f(\alpha_{n]}),p} \quad \forall \alpha, \beta \in \Omega;$$

then $\pi(B_n) = \{\tilde{p} : p \in \pi(A_n)\}$;

(iv) for $n \geq -1$, *and* $\lambda, \mu \in \Omega_{n]}$ *satisfying* $f(\lambda) = f(\mu)$, *define* $e_{\lambda,\mu} \in B_n$ *by*

$$e_{\lambda,\mu}(\alpha, \beta) = \delta_{\alpha_{n]},\lambda} \delta_{\beta_{n]},\mu} \delta_{\alpha_{[n},\beta_{[n}} \quad \forall \alpha, \beta \in \Omega;$$

then

$$\{e_{\lambda,\mu} : \lambda, \mu \in \Omega_{n]}, f(\lambda) = f(\mu)\}$$

is a 'system of matrix units' for B_n;

(v) there exists an isomorphism ψ *of the tower (5.4.1) onto the tower (5.4.2) such that* $\psi(p) = \tilde{p} \ \ \forall p \in \pi(A_n), \ \ \forall n \geq -1$.

Proof: The proof of (i)–(iv) is routine computation. Assertion (v) is easily seen, via induction, to amount to the following statement which is indeed true:

If $A_0 \subset A_1$ and $B_0 \subseteq B_1$ are inclusions of finite-dimensional C^*-algebras, if $\theta : A_0 \to B_0$ is an isomorphism, and if there exists a bijection $\phi : \pi(A_1) \to \pi(B_1)$ such that the two inclusion matrices Λ_A and Λ_B satisfy $\Lambda_A(p, q) = \Lambda_B(\theta(p), \phi(q))$, then there exists an isomorphism $\tilde{\theta} : A_1 \to B_1$ which simultaneously extends both θ and ϕ. □

REMARK 5.4.2 *(1) If $x \in B_n$ and $a \in \mathcal{L}(\mathcal{H}_{n]})$ are related as in the definition of B_n, the assignment $x \mapsto a$ defines a faithful representation ρ_n of B_n on $\mathcal{H}_{n]}$; we shall consequently simply write $x(\alpha, \beta)$ for $(\rho_{n]}(x))(\alpha, \beta)$, whenever $x \in B_n, \alpha, \beta \in \Omega_{n]}$. Further, $\dim_{\mathbb{C}} \mathcal{H}_{n]} = \sum_{p \in \pi(A_n)} [\dim_{\mathbb{C}}(B_n \tilde{p})]^{\frac{1}{2}}$, and ρ_n contains each distinct irreducible representation of B_n exactly once.*

(2) Suppose $\psi, \psi' : A_n \to B_n$ are two isomorphisms which both map A_k to B_k and $p \in \pi(A_k)$ to $\tilde{p}$ for each $p \in \pi(A_k)$, for $0 \leq k \leq n$. Then $\theta = \psi^{-1} \circ \psi'$ is an automorphism of A_n with the following two properties: (a) $\theta(A_k) = A_k, 0 \leq k \leq n$; and (b) $\theta|_{Z(A_k)} = \mathrm{id}_{Z(A_k)}, 0 \leq k \leq n$.

On the other hand, any automorphism of $M_n(\mathbb{C})$ is inner; hence an automorphism of a finite-dimensional C^*-algebra is inner if and only if it acts trivially on the centre; this implies, by an easy induction argument, that an automorphism θ of A_n satisfies properties (a) and (b) of the preceding paragraph if and only if it has the form $\theta = \mathrm{Ad}_{u_0 u_1 \cdots u_n}$, where u_k is a unitary element of $A'_{k-1} \cap A_k$, for $0 \leq k \leq n$.

Suppose now that tr is a faithful tracial state on A_n, where we temporarily fix $n \geq 0$. For $-1 \leq k \leq n$, let $\vec{t}^{(k)} : \pi(A_k) \to [0,1]$ be the trace-vector corresponding to $\mathrm{tr}|_{A_k}$; thus $\vec{t}^{(k)}_p = \mathrm{tr}\, p_0$, where p_0 is any minimal projection in $A_k p$. In the following proposition, whose proof is another straightforward verification, we identify the A_k's with the corresponding B_k's of Proposition 5.4.1.

PROPOSITION 5.4.3 *(i) If $-1 \leq k \leq n$, and if $x \in B_k$, then $\mathrm{tr}\, x = \sum_{\alpha \in \Omega_{k]}} \vec{t}^{(k)}_{f(\alpha)}$ $x(\alpha, \alpha)$.*

(ii) If $-1 \leq k < n$, and if $x \in B_n$, then the tr-preserving conditional expectation E_{B_k} of B_n onto B_k is given by

$$(E_{B_k} x)(\alpha, \beta) = \delta_{\alpha_{[k}, \beta_{[k}} \sum_{\substack{\theta \in \Omega_{[k,n]} \\ s(\theta) = f(\alpha_{k]})}} \frac{\vec{t}^{(n)}_{f(\theta)}}{t^{(k)}_{f(\alpha_{k]})}} x(\alpha_{k]} \circ \theta, \beta_{k]} \circ \theta)$$

for all $\alpha, \beta \in \Omega_{n]}$, where we have used the symbol $\circ$ to denote concatenation of paths.

(iii) In particular, if $-1 \leq k \leq n$, and if $e_{\lambda,\mu}$ is as in Proposition 5.4.1(iv), then

$$E_{B_k}(e_{\lambda,\mu}) = \delta_{\lambda_{[k}, \mu_{[k}} \frac{\vec{t}^{(n)}_{f(\lambda)}}{t^{(k)}_{f(\lambda_{k]})}} e_{\lambda_{k]}, \mu_{k]}}. \qquad \square$$

We conclude this section by discussing the above path-algebra formulation in the important special case when the tower (5.4.1) (equivalently, (5.4.2)) is the tower of the basic construction applied to the initial connected inclusion $A_0 \subseteq A_1$, with respect to the Markov trace. In this case, suppose Λ (resp., Ω_0) denotes the inclusion matrix (resp., the set of oriented edges in the Bratteli diagram) for $A_0 \subseteq A_1$. Let us inductively identify $p \in \pi(A_n)$

with $J_{A_{n+1}} p J_{A_{n+1}} \in \pi(A_{n+2})$. Further, if $\alpha \in \Omega_0$, let us write $\tilde{\alpha}$ for the same edge with the opposite orientation, and let us write $\tilde{\Omega_0} = \{\tilde{\alpha} : \alpha \in \Omega_0\}$. It then follows from Lemma 5.3.3 that if Λ_n denotes the inclusion matrix for $A_n \subseteq A_{n+1}$, then (Λ_n, Ω_n) is (Λ, Ω_0) or $(\Lambda', \tilde{\Omega_0})$ according as n is even or odd.

Let $\bar{t}^{(n)} : \pi(A_n) \to [0,1]$ denote the trace-vector corresponding to the restriction to A_n of tr, the unique consistently defined trace on $A_\infty = \bigcup_k A_k$. Then, with our identification of $\pi(A_{2n})$ (resp., $\pi(A_{2n+1})$) with $\pi(A_0)$ (resp., $\pi(A_1)$), it follows that

$$\bar{t}^{(2n+1)} = \tau^{-n} \bar{t}^{(1)}, \bar{t}^{(2n)} = \tau^{-n} \bar{t}^{(0)}, \bar{t}^{(0)} = \Lambda \bar{t}^{(1)},$$

where $\tau = ||\Lambda||^{-2}$.

Again, in the following result, which is easily verified, we identify A_n above with the corresponding B_n of Proposition 5.4.1.

PROPOSITION 5.4.4 *With the foregoing notation, a candidate for the projection $e_{n+1} \in A'_{n-1} \cap A_{n+1}$ is given by*

$$e_{n+1}(\alpha, \beta)$$

$$= \delta_{\alpha_{n-1]}, \beta_{n-1]}} \delta_{\alpha_{[n+1}, \beta_{[n+1}} \delta_{\alpha_{[n,n+1]}, \tilde{\alpha}_{[n-1,n]}} \delta_{\beta_{[n,n+1]}, \tilde{\beta}_{[n-1,n]}} \frac{\sqrt{\bar{t}^{(n)}_{f(\alpha_{n]})} \bar{t}^{(n)}_{f(\beta_{n]})}}}{\bar{t}^{(n-1)}_{f(\alpha_{n-1]})}}. \quad (5.4.3)$$

□

5.5 The biunitarity condition

In this section, we discuss the important re-formulation, due to Ocneanu, of the commuting square condition, as a *biunitarity* condition. Before we get to that, it would be good to make a slight digression concerning matrix units.

Recall that if e_{ij} is the $n \times n$ matrix with 1 in the (i,j) position and 0's elsewhere, then $\{e_{ij} : 1 \leq i, j \leq n\}$ is called a (or, sometimes, *the*) standard system of matrix units for $M_n(\mathbb{C})$; more generally, if A is a I_n factor, a set of matrix units for A is any set of the form $\{\psi^{-1}(e_{ij}) : 1 \leq i, j \leq n\}$, where $\psi : A \to M_n(\mathbb{C})$ is any isomorphism. Slightly more generally, if A is a finite-dimensional C^*-algebra, we shall refer to a union of sets of matrix units for each central summand of A as a system of matrix units for A. (See Proposition 5.4.1(iv).) Finally if we are given a tower of finite-dimensional C^*-algebras, say $A_0 \subseteq A_1 \subseteq \cdots \subseteq A_n$ as in the last section, we would like to work with a system of matrix units for A_n which is *compatible with the tower* in a natural fashion.

For this, notice that, in the notation of Proposition 5.4.1, the set $\{e_{\lambda,\mu} : \lambda, \mu \in \Omega_{n]}, f(\lambda) = f(\mu)\}$ is a system of matrix units for A_n with the following property: whenever $-1 \leq k \leq l \leq n, \lambda, \mu \in \Omega_{[k,l]}, s(\lambda) = s(\mu), f(\lambda) = f(\mu)$, if we define

$$e_{\lambda,\mu} = \sum_{\substack{\alpha \in \Omega_{k]}, \beta \in \Omega_{[l} \\ f(\alpha)=s(\lambda), s(\beta)=f(\lambda)}} e_{\alpha\circ\lambda\circ\beta, \alpha\circ\mu\circ\beta},$$

then $\{e_{\lambda,\mu} : \lambda, \mu \in \Omega_{[k,l]}, s(\lambda) = s(\mu), f(\lambda) = f(\mu)\}$ is a set of matrix units for $A'_k \cap A_l$.

DEFINITION 5.5.1 *If $A_0 \subseteq A_1 \subseteq \cdots \subseteq A_n$ is a tower of finite-dimensional C^*-algebras, we shall use the expression* a system of matrix units for A_n which is compatible with the tower *above, to denote any set of the form $\psi^{-1}(\{e_{\lambda,\mu} : \lambda, \mu \in \Omega_{n]}, f(\lambda) = f(\mu)\})$, where ψ and $\{e_{\lambda,\mu} : \lambda, \mu \in \Omega_{n]}, f(\lambda) = f(\mu)\}$ are as in Proposition 5.4.1(v) and (iv).*

Thus a consequence of Remark 5.4.2(2) is that any system of matrix units for A_n which is compatible with the tower $\{A_k : 0 \leq k \leq n\}$ can be obtained from any other by an inner automorphism of A_n of the form $\mathrm{Ad}_{u_0 u_1 \cdots u_n}$, where u_k is a unitary element of $A'_{k-1} \cap A_k$, for $0 \leq k \leq n$.

If $\mathcal{B}_{n]} = \{p_{\lambda,\mu} : \lambda, \mu \in \Omega_{n]}, s(\lambda) = s(\mu), f(\lambda) = f(\mu)\}$ is a system of matrix units for A_n which is compatible with the tower $A_0 \subseteq \cdots \subseteq A_n$, and if $\mathcal{B}_{[k,l]} = \{p_{\lambda,\mu} : \lambda, \mu \in \Omega_{[k,l]}, s(\lambda) = s(\mu), f(\lambda) = f(\mu)\}$ is the system of matrix units for $B'_k \cap B_l$ as above, we shall say that the system $\mathcal{B}_{[k,l]}$ extends to the system $\mathcal{B}_{n]}$, or that the system $\mathcal{B}_{n]}$ restricts to $\mathcal{B}_{[k,l]}$.

We are now ready for the biunitarity condition. (In the following proposition, we shall use the following notation: if $A_0 \subseteq A_1 \subseteq \cdots \subseteq A_n$ is a tower of finite-dimensional C^*-algebras, we write $\Omega_{(A_0;A_1;\cdots;A_n)}$ to denote the set of oriented paths which was denoted by $\Omega_{[0,n]}$ in the last section. We shall also use the convention that the symbol $\alpha \circ \beta$ will be used to denote concatenation of the paths α and β, and it will be tacitly assumed that if this symbol is used, then necessarily $s(\beta) = f(\alpha)$.)

PROPOSITION 5.5.2 *Let*

$$\begin{array}{ccc} B_0 & \subseteq & B_1 \\ \cup & & \cup \\ A_0 & \subseteq & A_1 \end{array} \tag{5.5.1}$$

be a square of algebras. Denote an element of $\Omega_{(\mathbb{C};A_0;A_1;B_1)}$ (resp., $\Omega_{(\mathbb{C};A_0;B_0;B_1)}$) by a triple $\alpha = (\alpha_0, \alpha_1, \alpha_2)$ (resp., $\beta = (\beta_0, \beta_1, \beta_2)$) where $\alpha_0, \beta_0 \in \Omega_{(\mathbb{C};A_0)}$, etc., so that, for instance, $\beta = \beta_0 \circ \beta_1 \circ \beta_2$.)

(a) Let $\{p_{\alpha,\alpha'} : \alpha, \alpha' \in \Omega_{(\mathbb{C};A_0;A_1;B_1)}, f(\alpha) = f(\alpha')\}$ (resp., $\{q_{\beta,\beta'} : \beta, \beta' \in \Omega_{(\mathbb{C};A_0;B_0;B_1)}, f(\beta) = f(\beta')\}$) be a system of matrix units for B_1 which is compatible with the tower $\mathbb{C} \subseteq A_0 \subseteq A_1 \subseteq B_1$ (resp., $\mathbb{C} \subseteq A_0 \subseteq B_0 \subseteq B_1$). Then there exists a matrix $U = ((u^\beta_\alpha))$, with columns (resp., rows) indexed by $\Omega_{(A_0;A_1;B_1)}$ (resp., $\Omega_{(A_0;B_0;B_1)}$), satisfying the following conditions:

(i) $u^\beta_\alpha = 0$ unless $s(\alpha) = s(\beta), f(\alpha) = f(\beta)$;

(ii) for all $\alpha, \alpha' \in \Omega_{(\mathbb{C};A_0;A_1;B_1)}$, *we have*

$$p_{\alpha,\alpha'} = \sum_{\beta,\beta' \in \Omega_{(A_0;B_0;B_1)}} u^{\beta}_{\alpha_1 \circ \alpha_2} \overline{u^{\beta'}_{\alpha'_1 \circ \alpha'_2}} q_{\alpha_0 \circ \beta, \alpha'_0 \circ \beta'};$$

(iii) U *is unitary.*

(b) If U' *is another matrix satisfying (a)(i)–(iii), then there exist complex scalars* $\omega_p, p \in \pi(B_1)$, *of unit modulus, such that* $(u')^{\beta}_{\alpha} = \omega_{f(\alpha)} u^{\beta}_{\alpha}$.

(c) Suppose tr *is a faithful tracial state on* B_1, *and suppose* $\bar{t}^C : \pi(C) \to [0,1]$ *is the trace vector corresponding to* $\mathrm{tr}|_C$, *for* $C \in \{A_0, A_1, B_0, B_1\}$; *define a matrix* V *with columns (resp., rows) indexed by* $\tilde{\Omega}_{(A_0;A_1)} \circ \Omega_{(A_0;B_0)} = \{\tilde{\alpha}_1 \circ \beta_1 : \alpha_1 \in \Omega_{(A_0;A_1)}, \beta_1 \in \Omega_{(A_0;B_0)}\}$ *(resp.,* $\Omega_{(A_1;B_1)} \circ \tilde{\Omega}_{(B_0;B_1)} = \{\alpha_2 \circ \tilde{\beta}_2 : \alpha_2 \in \Omega_{(A_1;B_1)}, \beta_2 \in \Omega_{(B_0;B_1)}\}$*) as follows:*

$$v^{\alpha_2 \circ \tilde{\beta}_2}_{\tilde{\alpha}_1 \circ \beta_1} = \left\{ \begin{array}{ll} \sqrt{\frac{\bar{t}^{A_0}_{s(\alpha_1)} \bar{t}^{B_1}_{f(\beta_2)}}{\bar{t}^{A_1}_{f(\alpha_1)} \bar{t}^{B_0}_{f(\beta_1)}}} \, \overline{u}^{\beta_1 \circ \beta_2}_{\alpha_1 \circ \alpha_2} & \text{if } (s(\alpha_2), s(\beta_2)) = (f(\alpha_1), f(\beta_1)), \\ 0 & \text{otherwise;} \end{array} \right\} \quad (5.5.2)$$

then (5.5.1) is a commuting square with respect to tr *if and only if* V *is an isometry – i.e., the columns of* V *constitute an orthonormal set of vectors.*

Proof: (a) Let ψ (resp., ψ') denote an isomorphism of B_1 onto the path-algebra model for B_1 coming from the tower $A_0 \subseteq A_1 \subseteq B_1$ (resp., $A_0 \subseteq B_0 \subseteq B_1$), as in Proposition 5.4.1(v), such that $\psi(p_{\alpha,\alpha'}) = e_{\alpha,\alpha'} \ \forall \alpha, \alpha' \in \Omega_{(\mathbb{C};A_0;A_1;B_1)}$ (resp., $\psi(q_{\beta,\beta'}) = e_{\beta,\beta'} \ \forall \beta, \beta' \in \Omega_{(\mathbb{C};A_0;B_0;B_1)}$); and let ρ_0 (resp., ρ'_0) denote the representation of $\psi(B_1)$ (resp., $\psi'(B_1)$) on a Hilbert space $\mathcal{H}$ (resp., $\mathcal{H}'$) with orthonormal basis $\{\xi_\alpha : \alpha \in \Omega_{(\mathbb{C};A_0;A_1;B_1)}\}$ (resp., $\{\eta_\beta : \beta \in \Omega_{(\mathbb{C};A_0;B_0;B_1)}\}$), as in Remark 5.4.2(1).

Consider the representations $\rho = \rho_0 \circ \psi$ and $\rho' = \rho'_0 \circ \psi'$ of B_1. It follows from Remark 5.4.2(1) that these representations are unitary equivalent. Let $U : \mathcal{H} \to \mathcal{H}'$ be a unitary operator which intertwines ρ and ρ'. The fact that U intertwines the restrictions to A_0 of the representations ρ and ρ', is easily seen to imply that there exist scalars $u^{\beta}_{\alpha}, \alpha \in \Omega_{(A_0;A_1;B_1)}, \beta \in \Omega_{(A_0;B_0;B_1)}(s(\alpha), f(\alpha)) = (s(\beta), f(\beta))$ (which are essentially just the matrix coefficients of U with respect to the given bases), such that

$$U\xi_\alpha = \sum_{\substack{\beta \in \Omega_{(A_0;B_0;B_1)} \\ s(\beta) = f(\alpha_0), f(\beta) = f(\alpha_2)}} u^{\beta}_{\alpha_1 \circ \alpha_2} \eta_{\alpha_0 \circ \beta}$$

whenever $\alpha = \alpha_0 \circ \alpha_1 \circ \alpha_2 \in \Omega_{(\mathbb{C};A_0;A_1;B_1)}$.

Since $\rho(p_{\alpha,\alpha'}) = \langle\cdot,\xi_{\alpha'}\rangle\xi_\alpha$ and $\rho'(q_{\beta,\beta'}) = \langle\cdot,\eta_{\beta'}\rangle\eta_\beta$, it follows that

$$\begin{aligned}\rho'(p_{\alpha,\alpha'}) &= U\rho(p_{\alpha,\alpha'})U^* \\ &= \sum_{\beta,\beta'\in\Omega_{(A_0;B_0;B_1)}} u^\beta_{\alpha_1\circ\alpha_2}\overline{u^{\beta'}_{\alpha'_1\circ\alpha'_2}}\rho'(q_{\alpha_0\circ\beta,\alpha'_0\circ\beta'}),\end{aligned}$$

thereby proving (a).

(b) In the notation of the proof of (a), notice that if U' is another matrix as in (a), and if we use the same symbol for the associated unitary operator from $\mathcal{H}$ to $\mathcal{H}'$ which is easily seen to necessarily intertwine the representations ρ and ρ', then $U'^*U \in \rho(B_1)'$; but it is a consequence of Remark 5.4.2(1) that (the representation ρ is 'multiplicity free' and hence) $\rho(B_1)' = \rho(Z(B_1))$, which is seen to imply (b).

(c) If $\alpha,\alpha' \in \Omega_{(\mathbb{C};A_0;A_1)}$ satisfy $f(\alpha) = f(\alpha')$, define

$$p_{\alpha,\alpha'} = \sum_{\alpha_2\in\Omega_{(A_1;B_1)}} p_{\alpha\circ\alpha_2,\alpha'\circ\alpha_2}.$$

Then, according to the remarks preceding Definition 5.5.1, it is the case that the set $\{p_{\alpha,\alpha'} : \alpha,\alpha' \in \Omega_{(\mathbb{C};A_0;A_1)}, f(\alpha) = f(\alpha')\}$ is a system of matrix units for A_1. Let $\{q_{\beta,\beta'} : \beta,\beta' \in \Omega_{(\mathbb{C};A_0;B_0)}, f(\beta) = f(\beta')\}$ be the system of matrix units for B_0 obtained in a similar fashion.

Then for fixed $\alpha,\alpha' \in \Omega_{(\mathbb{C};A_0;A_1)}$ satisfying $f(\alpha) = f(\alpha')$, we have

$$p_{\alpha,\alpha'} = \sum_{\substack{\alpha_2\in\Omega_{(A_1;B_1)}\\ \beta,\beta'\in\Omega_{(A_0;B_0;B_1)}}} u^\beta_{\alpha_1\circ\alpha_2}\overline{u^{\beta'}_{\alpha'_1\circ\alpha_2}}q_{\alpha_0\circ\beta,\alpha'_0\circ\beta'}$$

and hence, by Proposition 5.4.3(iii),

$$E_{B_0}(p_{\alpha,\alpha'}) = \sum_{\substack{\alpha_2\in\Omega_{(A_1;B_1)}\\ \beta,\beta'\in\Omega_{(A_0;B_0;B_1)}}} u^\beta_{\alpha_1\circ\alpha_2}\overline{u^{\beta'}_{\alpha'_1\circ\alpha_2}}\delta_{\beta_2,\beta'_2}\frac{\bar{t}^{B_1}_{f(\beta_2)}}{\bar{t}^{B_0}_{f(\beta_1)}}q_{\alpha_0\circ\beta_1,\alpha'_0\circ\beta'_1};$$

it follows now from the definition of the matrix V that

$$E_{B_0}(p_{\alpha,\alpha'}) = \frac{\bar{t}^{A_1}_{f(\alpha)}}{\sqrt{\bar{t}^{A_0}_{f(\alpha_0)}\bar{t}^{A_0}_{f(\alpha'_0)}}}\sum_{\beta_1,\beta'_1\in\Omega_{(B_0;B_1)}}(V^*V)^{\tilde{\alpha}_1\circ\beta_1}_{\tilde{\alpha}'_1\circ\beta'_1}q_{\alpha_0\circ\beta_1,\alpha'_0\circ\beta'_1}. \quad (5.5.3)$$

On the other hand, since $p_{\alpha_0,\alpha'_0} = q_{\alpha_0,\alpha'_0}$ whenever $\alpha_0,\alpha'_0 \in \Omega_{(\mathbb{C};A_0)}$ satisfy $f(\alpha_0) = f(\alpha'_0)$, it follows from another application of Proposition 5.4.3(iii) that

$$\begin{aligned}E_{A_0}(p_{\alpha,\alpha'}) &= \delta_{\alpha_1,\alpha'_1}\frac{\bar{t}^{A_1}_{f(\alpha)}}{\bar{t}^{A_0}_{f(\alpha_0)}}p_{\alpha_0,\alpha'_0} \\ &= \delta_{\alpha_1,\alpha'_1}\frac{\bar{t}^{A_1}_{f(\alpha)}}{\bar{t}^{A_0}_{f(\alpha_0)}}\sum_{\beta_1\in\Omega_{(A_0;B_0)}} q_{\alpha_0\circ\beta_1,\alpha'_0\circ\beta_1}. \quad (5.5.4)\end{aligned}$$

Since the $q_{\alpha_0 \circ \beta_1, \alpha_0' \circ \beta_1'}$'s are linearly independent, and since the $p_{\alpha,\alpha'}$'s form a basis for A_1, the assertion (c) follows from equations (5.5.3) and (5.5.4). □

REMARK 5.5.3 *(1) It must be clear that (5.5.1) is a symmetric commuting square if and only if the matrix V of Proposition 5.5.2(c) is unitary; this is the reason for referring to the conclusion of Proposition 5.5.2(c) as a biunitarity condition.*

(2) It must also be clear that if G, H, K and L are given rectangular matrices with non-negative integral entries, and if these satisfy the obvious consistency condition $GL = KH$, then a necessary and sufficient condition for the existence of a commuting square of algebras such as (5.5.1) with the inclusion matrices of $A_0 \subseteq A_1, B_0 \subseteq B_1, A_0 \subseteq B_0$ and $A_1 \subseteq B_1$ being G, H, K and L respectively, is the existence of a unitary matrix U as in Proposition 5.5.2 (with $\Omega_{(A_0;A_1)}$ being the set of oriented edges in a bipartite graph with 'adjacency' relations described by G, etc.), such that the corresponding V is an isometry.

(3) In the passage from the square (5.5.1) to the 'biunitary' matrix U, it was necessary to fix two systems of matrix units for B_1 compatible with the towers $\mathbb{C} \subseteq A_0 \subseteq A_1 \subseteq B_1$ and $\mathbb{C} \subseteq A_0 \subseteq B_0 \subseteq B_1$ respectively. It follows from Remark 5.4.2(2) that the 'isomorphism class' of a commuting square is described by an equivalence class of 'biunitary' matrices U as in Proposition 5.5.2, where two such matrices, say U and $\tilde{U}$, are equivalent if $\tilde{U} = W_2 U W_1$, where W_1 (resp., W_2) is a unitary matrix with rows and columns indexed by $\Omega_{(A_0;A_1;B_1)}$ (resp., $\Omega_{(A_0;B_0;B_1)}$) such that the matrix W_1 (resp., W_2) has the form $(W_1)^{\alpha_1 \circ \alpha_2}_{\alpha_1' \circ \alpha_2'} = \delta_{(s(\alpha_1),f(\alpha_1),f(\alpha_2)),(s(\alpha_1'),f(\alpha_1'),f(\alpha_2'))} (a_1)^{\alpha_1}_{\alpha_1'} (a_2)^{\alpha_2}_{\alpha_2'}$ (resp., $(W_2)^{\beta_1 \circ \beta_2}_{\beta_1' \circ \beta_2'} = \delta_{(s(\beta_1),f(\beta_1),f(\beta_2)),(s(\beta_1'),f(\beta_1'),f(\beta_2'))} (b_1)^{\beta_1}_{\beta_1'} (b_2)^{\beta_2}_{\beta_2'}$) where a_1, a_2, b_1, b_2 are suitably indexed unitary matrices.

(4) The passage $U \mapsto V$ respects equivalences of biunitary matrices (which describe symmetric commuting squares with respect to the Markov trace); i.e., suppose $U, \tilde{U}, W_1, W_2, a_1, a_2, b_1, b_2$ are as above; suppose $\tilde{V}$ is obtained from $\tilde{U}$ by the same prescription by which V was obtained from U, then it is easy to see that the biunitary matrices $\tilde{V}$ and V are also equivalent; to be brutally explicit, one has $\tilde{V} = \tilde{W}_2 V \tilde{W}_1$, where

$$\begin{aligned}
(\tilde{W}_2)^{\alpha_2 \circ \tilde{\beta}_2}_{\alpha_2' \circ \tilde{\beta}_2'} &= \delta_{(s(\alpha_2),f(\alpha_2),s(\beta_2)),(s(\alpha_2'),f(\alpha_2'),s(\beta_2'))} (\overline{a_2})^{\alpha_2}_{\alpha_2'} (\overline{b_2})^{\beta_2}_{\beta_2'}, \\
(\tilde{W}_1)^{\tilde{\alpha_1} \circ \beta_1}_{\tilde{\alpha_1}' \circ \beta_1'} &= \delta_{(s(\alpha_1),f(\alpha_1),f(\beta_1)),(s(\alpha_1'),f(\alpha_1'),f(\beta_1'))} (\overline{a_1})^{\alpha_1}_{\alpha_1'} (\overline{b_1})^{\beta_1}_{\beta_1'}.
\end{aligned}$$

We shall say that *the commuting square (5.5.1) is described by the biunitary matrix U* if the commuting square and the matrix are related as in Proposition 5.5.2. Thus, this happens precisely when it is possible to find two systems of matrix units $\{p_{\alpha,\alpha'}\}$ and $\{q_{\beta,\beta'}\}$ for B_1, compatible with respect to the towers $A_0 \subseteq A_1 \subseteq B_1$ and $A_0 \subseteq B_0 \subseteq B_1$ respectively, which are related by the matrix U as in Proposition 5.5.2(a).

We now want to consider the case where (5.5.1) is a symmetric commuting square with respect to the Markov trace, and consider the 'dual' (symmetric) commuting square obtained, as in Lemma 5.3.3, from the basic construction.

PROPOSITION 5.5.4 *Let A_i, B_i, e be as in Lemma 5.3.3. Suppose in addition that (5.3.2) is a symmetric commuting square. Suppose this symmetric commuting square is described by the biunitary matrix U as in Proposition 5.5.2, and suppose V is the matrix associated with U as in Proposition 5.5.2(c).*

(a)

$$\begin{array}{ccc} B_1 & \overset{H'}{\subset} & B_2 \\ {\scriptstyle L}\cup & & \cup{\scriptstyle K} \\ A_1 & \overset{G'}{\subset} & A_2, \end{array} \tag{5.5.5}$$

is also a symmetric commuting square, and it is described by the biunitary matrix $U_{[1,2]} = V$;

(b) further,

$$\begin{array}{ccc} B_0 & \overset{HH'}{\subset} & B_2 \\ {\scriptstyle K}\cup & & \cup{\scriptstyle K} \\ A_0 & \overset{GG'}{\subset} & A_2 \end{array} \tag{5.5.6}$$

is also a symmetric commuting square,and it is described by the biunitary matrix $U_{[0,2]}$ defined by

$$(u_{[0,2]})^{\kappa_1 \circ \beta_1 \circ \tilde{\beta}_2}_{\alpha_1 \circ \tilde{\alpha}_2 \circ \kappa_2} = \sum_{\lambda \in \Omega_{(A_1;B_1)}} u^{\kappa_1 \circ \beta_1}_{\alpha_1 \circ \lambda} v^{\lambda \circ \tilde{\beta}_2}_{\tilde{\alpha}_2 \circ \kappa_2}. \tag{5.5.7}$$

Proof: It follows already from Lemma 5.3.3 that the inclusions in the squares (5.5.5) and (5.5.6) are as indicated, and that (5.5.5), and consequently (5.5.6), is a commuting square with respect to the Markov trace. Further, it is obvious that the symmetry of the commuting square (5.3.2) implies that of the commuting squares (5.5.5) and (5.5.6). So we only need to prove the assertions concerning biunitary matrices.

We shall use the symbols $\alpha_0, \alpha_1, \tilde{\alpha}_2, \beta_1, \tilde{\beta}_2, \kappa_1, \lambda$ and κ_2 (or 'primed' versions of them) respectively, to denote the typical oriented path in $\Omega_{(\mathbf{C};A_0)}$, $\Omega_{(A_0;A_1)}$, $\Omega_{(A_1;A_2)}$, $\Omega_{(B_0;B_1)}$, $\Omega_{(B_1;B_2)}$, $\Omega_{(A_0;B_0)}$, $\Omega_{(A_1;B_1)}$ and $\Omega_{(A_2;B_2)}$, with the understanding that α_2 (resp., β_2) is an oriented path in $\Omega_{(A_0;A_1)}$ (resp., $\Omega_{(B_0;B_1)}$).

Suppose $\{r_{\alpha_0 \circ \alpha_1 \circ \lambda, \alpha_0' \circ \alpha_1' \circ \lambda'}\}$ and $\{q_{\alpha_0 \circ \kappa_1 \circ \beta_1, \alpha_0' \circ \kappa_1' \circ \beta_1'}\}$ denote systems of matrix units for B_1 compatible with the towers $\mathbf{C} \subseteq A_0 \subseteq A_1 \subseteq B_1$ and $\mathbf{C} \subseteq A_0 \subseteq B_0 \subseteq B_1$ respectively, and that the biunitary matrix U is related to the commuting square (5.3.2) via these systems of matrix units; thus,

$$r_{\alpha_0 \circ \alpha_1 \circ \lambda, \alpha_0' \circ \alpha_1' \circ \lambda'} = \sum_{\kappa_1 \circ \beta_1, \kappa_1' \circ \beta_1'} u^{\kappa_1 \circ \beta_1}_{\alpha_1 \circ \lambda} \overline{u}^{\kappa_1' \circ \beta_1'}_{\alpha_1' \circ \lambda'} q_{\alpha_0 \circ \kappa_1 \circ \beta_1, \alpha_0' \circ \kappa_1' \circ \beta_1'}. \tag{5.5.8}$$

It is clear from Proposition 5.4.4 and Lemma 5.3.1 that there exists an isomorphism, call it ψ_3, of B_2 onto the path-algebra model for B_2 coming from the tower $\mathbf{C} \subseteq A_0 \subseteq B_0 \subseteq B_1 \subseteq B_2$, such that ψ_3 maps the system

$\{q_{\alpha_0\circ\kappa_1\circ\beta_1,\alpha'_0\circ\kappa'_1\circ\beta'_1}\}$ of matrix units for B_1 onto the standard system of matrix units for the path algebra given by $\{e_{\alpha_0\circ\kappa_1\circ\beta_1,\alpha'_0\circ\kappa'_1\circ\beta'_1}\}$, and such that

$$\psi_3(e) = \sum_{\beta_1,\beta'_1,s(\beta_1)=s(\beta'_1)} \frac{\sqrt{\bar{t}^{B_1}_{f(\beta_1)}\bar{t}^{B_1}_{f(\beta'_1)}}}{\bar{t}^{B_0}_{s(\beta_1)}} e_{\beta_1\circ\tilde{\beta}_1,\beta'_1\circ\tilde{\beta'}_1}, \tag{5.5.9}$$

where, as before, we write $\bar{t}^C : \pi(C) \to [0,1]$ for the trace-vector corresponding to $\mathrm{tr}|_C$, for $C \in \{A_i, B_i : i = 0,1,2\}$. Let $\mathcal{B}_3 = \{q_{\alpha_0\circ\kappa_1\circ\beta_1\circ\tilde{\beta}_2,\alpha'_0\circ\kappa'_1\circ\beta'_1\circ\tilde{\beta'}_2}\}$ be the system of matrix units for B_2 (which is compatible with the tower $\mathbb{C} \subseteq A_0 \subseteq B_0 \subseteq B_1 \subseteq B_2$) obtained as the inverse image under ψ_3 of the standard system $\{e_{\alpha_0\circ\kappa_1\circ\beta_1\circ\tilde{\beta}_2,\alpha'_0\circ\kappa'_1\circ\beta'_1\circ\tilde{\beta'}_2}\}$ of matrix units for the path-algebra model. (It should be clear that this notation is justified – in the sense that this system of matrix units does indeed extend the system $\{q_{\alpha_0\circ\kappa_1\circ\beta_1,\alpha'_0\circ\kappa'_1\circ\beta'_1}\}$ chosen earlier.)

Similarly there exists an isomorphism ψ^0_1 of A_2 onto the path-algebra model for A_2 coming from the tower $\mathbb{C} \subseteq A_0 \subseteq A_1 \subseteq A_2$, such that $\psi^0_1(r_{\alpha_0\circ\alpha_1,\alpha'_0\circ\alpha'_1}) = e_{\alpha_0\circ\alpha_1,\alpha'_0\circ\alpha'_1}$ and such that

$$\psi^0_1(e) = \sum_{\alpha_1,\alpha'_1,s(\alpha_1)=s(\alpha'_1)} \frac{\sqrt{\bar{t}^{A_1}_{f(\alpha_1)}\bar{t}^{A_1}_{f(\alpha'_1)}}}{\bar{t}^{A_0}_{s(\alpha_1)}} e_{\alpha_1\circ\tilde{\alpha}_1,\alpha'_1\circ\tilde{\alpha'}_1}. \tag{5.5.10}$$

Now let $\mathcal{B}_1 = \{p^0_{\alpha_0\circ\alpha_1\circ\tilde{\alpha}_2\circ\kappa_2,\alpha'_0\circ\alpha'_1\circ\tilde{\alpha'}_2\circ\kappa'_2}\}$ be any system of matrix units for B_2 which is compatible with the tower $\mathbb{C} \subseteq A_0 \subseteq A_1 \subseteq A_2 \subseteq B_2$ and which 'extends' the system $(\psi^0_1)^{-1}(\{e_{\alpha_0\circ\alpha_1\circ\tilde{\alpha}_2,\alpha'_0\circ\alpha'_1\circ\tilde{\alpha'}_2}\})$; let ψ_1 denote the isomorphism of B_2 onto the path-algebra model from the tower $\mathbb{C} \subseteq A_0 \subseteq A_1 \subseteq A_2 \subseteq B_2$ which maps $p^0_{\alpha_0\circ\alpha_1\circ\tilde{\alpha}_2\circ\kappa_2,\alpha'_0\circ\alpha'_1\circ\tilde{\alpha'}_2\circ\kappa'_2}$ onto $e_{\alpha_0\circ\alpha_1\circ\tilde{\alpha}_2\circ\kappa_2,\alpha'_0\circ\alpha'_1\circ\tilde{\alpha'}_2\circ\kappa'_2}$.

It should be clear that if we define

$$r_{\alpha_0\circ\alpha_1\circ\lambda\circ\tilde{\beta}_2,\alpha'_0\circ\alpha'_1\circ\lambda'\circ\tilde{\beta'}_2} = \sum_{\kappa_1\circ\beta_1,\kappa'_1\circ\beta'_1} u^{\kappa_1\circ\beta_1}_{\alpha_1\circ\lambda}\overline{u}^{\kappa'_1\circ\beta'_1}_{\alpha'_1\circ\lambda'} q_{\alpha_0\circ\kappa_1\circ\beta_1\circ\tilde{\beta}_2,\alpha'_0\circ\kappa'_1\circ\beta'_1\circ\tilde{\beta'}_2},$$

then $\mathcal{B}_2 = \{r_{\alpha_0\circ\alpha_1\circ\lambda\circ\tilde{\beta}_2,\alpha'_0\circ\alpha'_1\circ\lambda'\circ\tilde{\beta'}_2}\}$ is a system of matrix units for B_2 which is compatible with the tower $\mathbb{C} \subseteq A_0 \subseteq A_1 \subseteq B_1 \subseteq B_2$ and which extends the already chosen system of matrix units $\{r_{\alpha_0\circ\alpha_1\circ\lambda,\alpha'_0\circ\alpha'_1\circ\lambda'}\}$ for B_1. It should be obvious that we have the relations

$$q_{\alpha_0\circ\kappa_1\circ\beta_1\circ\tilde{\beta}_2,\alpha'_0\circ\kappa'_1\circ\beta'_1\circ\tilde{\beta'}_2} = \sum_{\alpha_1\circ\lambda,\alpha'_1\circ\lambda'} \overline{u}^{\kappa_1\circ\beta_1}_{\alpha_1\circ\lambda} u^{\kappa'_1\circ\beta'_1}_{\alpha'_1\circ\lambda'} r_{\alpha_0\circ\alpha_1\circ\lambda\circ\tilde{\beta}_2,\alpha'_0\circ\alpha'_1\circ\lambda'\circ\tilde{\beta'}_2}. \tag{5.5.11}$$

The definitions and equation (5.5.10) imply that

$$e = \sum_{\alpha_1,\alpha'_1,s(\alpha_1)=s(\alpha'_1)} \frac{\sqrt{\bar{t}^{A_1}_{f(\alpha_1)}\bar{t}^{A_1}_{f(\alpha'_1)}}}{\bar{t}^{A_0}_{s(\alpha_1)}} p^0_{\alpha_1\circ\tilde{\alpha}_1,\alpha'_1\circ\tilde{\alpha'}_1}. \tag{5.5.12}$$

Similarly, it follows from equation (5.5.9), equation (5.5.11), and the definition of the matrix V (in terms of U), that

$$\begin{aligned}
e &= \sum_{\beta_1,\beta'_1,s(\beta_1)=s(\beta'_1)} \frac{\sqrt{\bar{t}^{B_1}_{f(\beta_1)}\bar{t}^{B_1}_{f(\beta'_1)}}}{\bar{t}^{B_0}_{s(\beta_1)}} q_{\beta_1\circ\tilde{\beta}_1,\beta'_1\circ\tilde{\beta}'_1} \\
&= \sum_{\substack{\beta_1,\beta'_1 \\ s(\beta_1)=s(\beta'_1)}} \sum_{\substack{\alpha_0\circ\kappa_1 \\ f(\kappa_1)=s(\beta_1)}} \frac{\sqrt{\bar{t}^{B_1}_{f(\beta_1)}\bar{t}^{B_1}_{f(\beta'_1)}}}{\bar{t}^{B_0}_{s(\beta_1)}} q_{\alpha_0\circ\kappa_1\circ\beta_1\circ\tilde{\beta}_1,\alpha_0\circ\kappa_1\circ\beta'_1\circ\tilde{\beta}'_1} \\
&= \sum_{\substack{\alpha_0\circ\kappa_1\circ\beta_1,\beta'_1,\alpha_1\circ\lambda,\alpha'_1\circ\lambda' \\ s(\beta_1)=s(\beta'_1)}} \frac{\sqrt{\bar{t}^{B_1}_{f(\beta_1)}\bar{t}^{B_1}_{f(\beta'_1)}}}{\bar{t}^{B_0}_{s(\beta_1)}} \overline{u}^{\kappa_1\circ\beta_1}_{\alpha_1\circ\lambda} u^{\kappa_1\circ\beta'_1}_{\alpha'_1\circ\lambda'} r_{\alpha_0\circ\alpha_1\circ\lambda\circ\tilde{\beta}_1,\alpha_0\circ\alpha'_1\circ\lambda'\circ\tilde{\beta}'_1} \\
&= \sum_{\substack{\alpha_0\circ\kappa_1\circ\beta_1,\beta'_1,\alpha_1\circ\lambda,\alpha'_1\circ\lambda' \\ s(\beta_1)=s(\beta'_1)}} \frac{\sqrt{\bar{t}^{A_1}_{f(\alpha_1)}\bar{t}^{A_1}_{f(\alpha'_1)}}}{\bar{t}^{A_0}_{s(\alpha_1)}} v^{\lambda\circ\tilde{\beta}_1}_{\tilde{\alpha}_1\circ\kappa_1} \overline{v}^{\lambda'\circ\tilde{\beta}'_1}_{\tilde{\alpha}'_1\circ\kappa_1} r_{\alpha_0\circ\alpha_1\circ\lambda\circ\tilde{\beta}_1,\alpha_0\circ\alpha'_1\circ\lambda'\circ\tilde{\beta}'_1}.
\end{aligned}$$

Let $\mathcal{H}_1$ and $\mathcal{H}_2$ denote Hilbert spaces with orthonormal bases $\{\xi_\alpha : \alpha \in \Omega_{(\mathbf{C};A_0;A_1;A_2;B_2)}\}$ and $\{\eta_\gamma : \gamma \in \Omega_{(\mathbf{C};A_0;A_1;B_1;B_2)}\}$ respectively. Let ρ_1 (resp., ρ_2) denote the representation of B_2 on $\mathcal{H}_1$ (resp., $\mathcal{H}_2$) such that $\rho_1(p^0_{\alpha,\alpha'}) = \langle\cdot,\xi_{\alpha'}\rangle\xi_\alpha$ (resp., $\rho_2(r_{\gamma,\gamma'}) = \langle\cdot,\eta_{\gamma'}\rangle\eta_\gamma$). Consider the unitary operator $V : \mathcal{H}_1 \to \mathcal{H}_2$ defined by

$$V\xi_{\alpha_0\circ\alpha_1\circ\tilde{\alpha}_2\circ\kappa_2} = \sum_{\lambda\circ\tilde{\beta}_2} v^{\lambda\circ\tilde{\beta}_2}_{\tilde{\alpha}_2\circ\kappa_2}\eta_{\alpha_0\circ\alpha_1\circ\lambda\circ\tilde{\beta}_2}.$$

It is clear from the definitions that V intertwines the restrictions to A_1 of the representations ρ_1 and ρ_2, and that

$$\begin{aligned}
&V\rho_1(p^0_{\alpha_0\circ\alpha_1\circ\tilde{\alpha}_2\circ\kappa_2,\alpha'_0\circ\alpha'_1\circ\tilde{\alpha}'_2\circ\kappa'_2})V^* \\
&\quad= \sum_{\lambda\circ\tilde{\beta}_2,\lambda'\circ\tilde{\beta}'_2} v^{\lambda\circ\tilde{\beta}_2}_{\tilde{\alpha}_2\circ\kappa_2}\overline{v}^{\lambda'\circ\tilde{\beta}'_2}_{\tilde{\alpha}'_2\circ\kappa'_2}\rho_2(r_{\alpha_0\circ\alpha_1\circ\lambda\circ\tilde{\beta}_2,\alpha'_0\circ\alpha'_1\circ\lambda'\circ\tilde{\beta}'_2}). \qquad (5.5.13)
\end{aligned}$$

It follows quite easily from equation (5.5.12), equation (5.5.13) and the already obtained expression for e in terms of the $r_{\gamma,\gamma'}$'s (after making the change of variables (β_2,κ_2) to (β_1,κ_1)) that $V^*\rho_1(e)V = \rho_2(e)$. Since $A_2 = \langle A_1, e\rangle$, this means that the operator V intertwines the restrictions to A_2 of ρ_1 and ρ_2.

Suppose now that $((w^{\lambda\circ\tilde{\beta}_2}_{\tilde{\alpha}_2\circ\kappa_2}))$ is the biunitary matrix which describes the commuting square (5.5.5) with respect to the systems of matrix units $\{p^0_{\alpha,\alpha'}\}$ and $\{r_{\gamma,\gamma'}\}$; thus, if we define an operator $W : \mathcal{H}_1 \to \mathcal{H}_2$ by the prescription

$$W\xi_{\alpha_0\circ\alpha_1\circ\tilde{\alpha}_2\circ\kappa_2} = \sum_{\lambda\circ\tilde{\beta}_2} w^{\lambda\circ\tilde{\beta}_2}_{\tilde{\alpha}_2\circ\kappa_2}\eta_{\alpha_0\circ\alpha_1\circ\lambda\circ\tilde{\beta}_2},$$

then W is a unitary operator which intertwines the representations ρ_1 and ρ_2 of B_2.

It follows then that $W^*V \in \rho_1(A_2)'$; on the other hand, it is clear from the definitions that W^*V commutes with $\{\rho_1(f) : f \in Z(B_2)\}$; on the other hand, by Remark 5.4.2(1), we have $\rho_1(Z(B_2)) = \rho_1(B_2)'$; thus we find that $W^*V \in \rho_1(B_2)$, and that in fact $W^*V = \rho_1(u_0)$ for some unitary element $u_0 \in A_2' \cap B_2$. It is easy now to see that if we define $p_{\alpha,\alpha'} = u_0 p^0_{\alpha,\alpha'} u_0^*$, then $\{p_{\alpha,\alpha'}\}$ is a system of matrix units (which is compatible with the tower $\mathbb{C} \subseteq A_0 \subseteq A_1 \subseteq A_2 \subseteq B_2$) and that

$$p_{\alpha_0\circ\alpha_1\circ\tilde{\alpha_2}\circ\kappa_2,\alpha_0'\circ\alpha_1'\circ\tilde{\alpha'}_2\circ\kappa_2'} = \sum_{\lambda\circ\tilde{\beta_2},\lambda'\circ\tilde{\beta'}_2} v^{\lambda\circ\tilde{\beta_2}}_{\tilde{\alpha_2}\circ\kappa_2} \overline{v}^{\lambda'\circ\tilde{\beta'}_2}_{\tilde{\alpha'}_2\circ\kappa_2'} r_{\alpha_0\circ\alpha_1\circ\lambda\circ\tilde{\beta_2},\alpha_0'\circ\alpha_1'\circ\lambda'\circ\tilde{\beta'}_2}. \tag{5.5.14}$$

This proves (a), while a combination of equations (5.5.14) and (5.5.8) establishes (b). □

COROLLARY 5.5.5 *Suppose (5.3.2) is a symmetric commuting square with respect to the Markov trace. Let $\{B_n\}$ denote the tower of the basic cnstruction applied to the initial inclusion $B_0 \subseteq B_1$, with $e_{n+1} \in B_{n+1}$ implementing the conditional expectation of B_n onto B_{n-1}. Inductively define $A_{n+1} = \langle A_n, e_{n+1}\rangle, n \geq 1$.*

(i) Then, for every $n \geq 0$,

$$\begin{array}{ccc} B_n & \subset & B_{n+1} \\ \cup & & \cup \\ A_n & \subset & A_{n+1} \end{array} \tag{5.5.15}$$

is also a symmetric commuting square,and it is described by the biunitary matrix $U_{[n,n+1]}$, where

$$U_{[n,n+1]} = \begin{cases} U & \text{if } n \text{ is even,} \\ V & \text{if } n \text{ is odd.} \end{cases}$$

(ii) For each $n \geq 0$, we have isomorphisms

$$\left(\begin{array}{ccc} A_n' \cap B_n & \subset & A_n' \cap B_{n+1} \\ \cup & & \cup \\ Z(A_n) & \subset & A_n' \cap A_{n+1} \end{array}\right) \cong \left(\begin{array}{ccc} A_{n+2}' \cap B_{n+2} & \subset & A_{n+2}' \cap B_{n+3} \\ \cup & & \cup \\ Z(A_{n+2}) & \subset & A_{n+2}' \cap A_{n+3} \end{array}\right).$$

Proof: (i) is an immediate consequence of Proposition 5.5.4(a) and the easily verifed fact that if we denote the passage $U \mapsto V$ by $V = V(U)$, then $V(V(U)) = U$.

(ii) By definition, we can find systems of matrix units $\{p^{(k)}_{\alpha,\alpha'} : \alpha, \alpha' \in \Omega_{(\mathbb{C};A_k;A_{k+1};B_{k+1})}\}$ and $\{q^{(k)}_{\beta,\beta'} : \beta, \beta' \in \Omega_{(\mathbb{C};A_k;B_k;B_{k+1})}\}$ for B_{k+1} (which are compatible with the towers $\mathbb{C} \subseteq A_k \subseteq A_{k+1} \subseteq B_{k+1}$ and $\mathbb{C} \subseteq A_k \subseteq B_k \subseteq B_{k+1}$ respectively, such that

$$p^{(k)}_{\alpha_0 \circ \alpha_k \circ \lambda_{k+1}, \alpha'_0 \circ \alpha'_k \circ \lambda'_{k+1}} = \sum_{\lambda_k \circ \beta_k, \lambda'_k \circ \beta'_k} (u_{[k,k+1]})^{\lambda_k \circ \beta_k}_{\alpha_k \circ \lambda_{k+1}} \overline{(u_{[k,k+1]})^{\lambda'_k \circ \beta'_k}_{\alpha'_k \circ \lambda'_{k+1}}} q^{(k)}_{\alpha_0 \circ \lambda_k \circ \beta_k, \alpha'_0 \circ \lambda'_k \circ \beta'_k}.$$

Since $\Omega_{(A_n;B_n;B_{n+1})} = \Omega_{(A_{n+2};B_{n+2};B_{n+3})}$, it follows from the path-algebra description that there exists a unique isomorphism $\psi : (A'_n \cap B_{n+1}) \to (A'_{n+2} \cap B_{n+3})$ such that $\psi(q^{(n)}_{\lambda_n \circ \beta_n, \lambda'_n \circ \beta'_n}) = q^{(n+2)}_{\lambda_n \circ \beta_n, \lambda'_n \circ \beta'_n}$. Since $U_{[n,n+1]} = U_{[n+2,n+3]}$, it follows that also $\psi(p^{(n)}_{\alpha_n \circ \lambda_{n+1}, \alpha'_n \circ \lambda'_{n+1}}) = p^{(n+2)}_{\alpha_n \circ \lambda_{n+1}, \alpha'_n \circ \lambda'_{n+1}}$, and so the map ψ implements the desired isomorphism. □

5.6 Canonical commuting squares

Assume, as usual, that $N \subseteq M$ is a finite-index inclusion of II_1 factors, that

$$N = M_{-1} \subseteq M = M_0 \subseteq M_1 \subseteq \cdots \subseteq M_n \subseteq \cdots$$

is the tower of the basic construction, and that e_{n+1} denotes, for $n \geq 0$, the projection in M_{n+1} which implements the conditional expectation of M_n onto M_{n-1}.

PROPOSITION 5.6.1 *Fix $n \geq 0$, and let* tr *denote the restriction, to $N' \cap M_{n+1}$, of $\mathrm{tr}_{M_{n+1}}$.*

(i) The following is a commuting square with respect to tr*:*

$$\begin{array}{ccc} N' \cap M_n & \subseteq & N' \cap M_{n+1} \\ \cup & & \cup \\ M' \cap M_n & \subseteq & M' \cap M_{n+1}. \end{array} \tag{5.6.1}$$

(ii) tr *is the unique Markov trace for the inclusion $(N' \cap M_n) \subseteq (N' \cap M_{n+1})$.*

Proof: Fix $x \in (M' \cap M_{n+1})$ (resp., $(N' \cap M_{n+1})$); then for arbitrary $a \in M$ (resp., N), an application of E_{M_n} to both sides of the equation $xa = ax$ yields $(E_{M_n} x)a = a(E_{M_n}(x))$. Since a was arbitrary, this means that $E_{M_n}(M' \cap M_{n+1}) \subseteq (M' \cap M_n)$ (resp., $E_{M_n}(N' \cap M_{n+1}) \subseteq (N' \cap M_n)$), which clearly implies the truth of (i).

As for (ii), note, as above, that the projection $e_{n+2} \in (N' \cap M_{n+2})$ implements the tr-preserving conditional expectation of $(N' \cap M_{n+1})$ onto $(N' \cap M_n)$ and that e_{n+2} satisfies the Markov property with respect to $(N' \cap M_{n+1})$ and $\mathrm{tr}_{M_{n+2}}|_{(N' \cap M_{n+2})}$. It follows now from Lemma 5.3.1(b) and Proposition 3.2.3 that tr is a Markov trace for the inclusion $(N' \cap M_n) \subseteq (N' \cap M_{n+1})$. On the other hand, as a result of the bimodule picture of the principal graph, this has already been seen to be a connected inclusion, and the proof is complete. □

To proceed with the analysis, it is desirable to take a closer look at Lemma 5.1.1 and its consequences. According to that lemma, if $N \subseteq M$ is a finite-index inclusion of II_1 factors, then we can find a subfactor N_{-1} of N such that M is isomorphic to the result of the basic construction for $N_{-1} \subseteq N$. However, the choice of N_{-1} is far from unique. For instance, if u is a unitary element of N, it should be clear that $uN_{-1}u^*$ works just as well as N_{-1}. (It is a fact – see [PP1] – that this turns out to be the only freedom available in the choice of N_{-1}; but we do not need that fact here.)

Exactly as we constructed the tower $\{M_n\}_{n\geq 1}$ of the basic construction, we may also construct a tunnel $\{N_{-n}\}_{n\geq 1}$ in such a way that every inclusion of neighbours in (5.6.2) (below) is an inclusion of II_1 factors with index equal to $[M : N]$, and such that any string of three equally spaced factors 'is a basic construction':

$$\cdots \subseteq N_{-n} \subseteq \cdots N_{-1} \subseteq N \subseteq M \subseteq M_1 \subseteq \cdots M_n \subseteq \cdots. \tag{5.6.2}$$

The point, however, is that while the 'tower' construction is canonical, the 'tunnel' construction is only 'semi-canonical' in the sense discussed in the last paragraph; thus, given N_{-n}, the subfactor N_{-n-1} is determined only up to an inner automorphism in N_{-n}.

In any case, suppose we have chosen a tunnel as in (5.6.2). A little thought should convince the reader that once we have fixed n, the map $x \mapsto J_M x^* J_M$ defines an anti-isomorphism of N_{-n} onto M_{n+1} which maps N_{-k} onto M_{k+1} for $1 \leq k \leq n$. Thus, we have, for each n, an anti-isomorphism (of commuting squares):

$$\begin{pmatrix} M' \cap M_n & \subseteq & M' \cap M_{n+1} \\ \cup & & \cup \\ M_1' \cap M_n & \subseteq & M_1' \cap M_{n+1} \end{pmatrix} \overset{\textit{anti-isom.}}{\cong} \begin{pmatrix} M \cap N'_{-(n-1)} & \subseteq & M \cap N'_{-n} \\ \cup & & \cup \\ N \cap N'_{-(n-1)} & \subseteq & N \cap N'_{-n} \end{pmatrix}. \tag{5.6.3}$$

It follows, thus, that independent of the manner in which the tunnel $\{N_{-n}\}_{n\geq 1}$ was constructed, the following grid of finite-dimensional C^* -algebras, equipped with consistently defined traces (given by the restrictions of tr_M) – or rather, the trace-preserving isomorphism-class of this grid – is an invariant of the inclusion $N \subseteq M$:

$$\begin{array}{ccccccccc} \mathbb{C} = M \cap M' & \subseteq & M \cap N' & \subseteq & M \cap N'_{-1} & \subseteq & M \cap N'_{-2} & \subseteq & \cdots \\ & & \cup & & \cup & & \cup & & \\ & & N \cap N' & \subseteq & N \cap N'_{-1} & \subseteq & N \cap N'_{-2} & \subseteq & \cdots . \end{array} \tag{5.6.4}$$

DEFINITION 5.6.2 *The trace-preserving isomorphism-class of the grid (5.6.4) of finite-dimensional C^*-algebras is called the standard invariant of the sub-*

factor $N \subseteq M$, and the individual squares in this grid are called the canonical commuting squares associated with the subfactor.

Notice, incidentally, that in view of the anti-isomorphisms given by (5.6.3), the standard invariant contains the data of the principal and dual graphs of the subfactor. (The latter (resp., the former) describes the Bratteli diagram of the tower contained in the top (resp., the bottom) row in the standard invariant.)

Further, it must be clear that the standard invariant would even be a complete invariant for the subfactor $N \subseteq M$, *provided* we could, by making a judicious set of choices, find a tunnel $\{N_{-n}\}$ with the (generating) property that $\cup_n (M \cap N'_{-n})$ is weakly dense in M and $\cup_n (N \cap N'_{-n})$ is weakly dense in N. Notice that an obvious necessary condition for this is that N and M be both isomorphic to the hyperfinite II_1 factor. (It is a fact that a finite-index subfactor of R is necessarily hyperfinite.)

It is a deep theorem due to Popa ([Pop6]) that a subfactor admits such a *generating tunnel* if and only if it satisfies a certain property he calls *strong amenability*. It follows from one of the equivalent ways of characterising this notion of strong amenability that any subfactor of *finite depth* – i.e., for which the principal (equivalently the dual) graph is a finite graph – is necessarily strongly amenable, and is hence characterised by its standard invariant.

It will be a good idea to consider the finite-depth case a little more closely. Suppose then that $N \subseteq M$ is a subfactor of finite depth. It then follows from the definition of the principal and dual graphs, that if n is large enough, then $(M'_1 \cap M_{n-1}) \subseteq (M'_1 \cap M_n) \subseteq (M'_1 \cap M_{n+1})$ is an instance of the basic construction, as is $(M' \cap M_{n-1}) \subseteq (M' \cap M_n) \subseteq (M' \cap M_{n+1})$, with the conditional expectation being implemented in either case by e_{n+1}. Thus, in view of Lemma 5.3.3 and Corollary 5.3.4, we may paraphrase the finite-depth case of Popa's theorem on amenable subfactors as follows:

THEOREM 5.6.3 *Let $N \subseteq M$ be a subfactor (of a II_1 factor) of finite depth. Then, for n sufficiently large,*

$$\begin{array}{ccc} M' \cap M_n & \subseteq & M' \cap M_{n+1} \\ \cup & & \cup \\ M'_1 \cap M_n & \subseteq & M'_1 \cap M_{n+1} \end{array} \tag{5.6.5}$$

is a symmetric commuting square. Further, the isomorphism class of this square of algebras is a complete invariant of the subfactor.

5.7 Ocneanu compactness

Suppose, in the opposite direction to Theorem 5.6.3, that we are given a symmetric commuting square (with respect to the Markov trace), say

$$\begin{array}{ccc} B_0 & \overset{H}{\subset} & B_1 \\ {\scriptstyle K}\cup & & \cup{\scriptstyle L} \\ A_0 & \overset{G}{\subset} & A_1, \end{array} \tag{5.7.1}$$

where the inclusions are assumed to be connected. Let

$$B_0 \subseteq B_1 \subseteq B_2 \subseteq B_3 \subseteq \cdots$$

be the tower of the basic construction (applied to the initial inclusion $B_0 \subseteq B_1$), where $B_{n+1} = \langle B_n, e_{n+1} \rangle$ for $n \geq 1$. Define $A_{n+1} = \langle A_n, e_{n+1} \rangle$ for $n \geq 1$. (Notice that, according to our conventions, the sequence $\{e_n\}$ now starts only with e_2.) It then follows from Lemma 5.3.3 and Corollary 5.3.4 that

$$\begin{array}{ccc} B_n & \overset{H_n}{\subset} & B_{n+1} \\ {\scriptstyle K_n}\cup & & \cup{\scriptstyle L_n} \\ A_n & \overset{G_n}{\subset} & A_{n+1} \end{array} \tag{5.7.2}$$

is a commuting square for each $n \geq 1$, with inclusion matrices as given, where (G_n, H_n, K_n, L_n) is (G, H, K, L) or (G', H', L, K) according as n is even or odd.

As has already been observed in §3.2, the algebra $\bigcup_n B_n$ admits a unique tracial state, and consequently has a copy R of the hyperfinite II_1 factor as its von Neumann algebra completion with respect to this trace; for the same reason, the weak closure of $\bigcup_n A_n$ is a subfactor R_0 of R.

Thus each symmetric commuting square of finite-dimensional C^*-algebras (with respect to the Markov trace, and such that the inclusions are connected) yields, by the above iterative procedure, a subfactor of the hyperfinite II_1 factor. The content of Theorem 5.6.3 is that if, in place of (5.3.2), we had started out with the canonical commuting square (5.6.5) associated with a finite-depth subfactor of the hyperfinite II_1 factor, then the subfactor R_0 obtained by the above prescription would have been 'anti-conjugate' to N in the sense that the inclusions $N \subseteq M$ and $R_0 \subseteq R$ would be anti-isomorphic.

Conversely, we can start from an arbitrary symmetric commuting square as above, obtain the subfactor $R_0 \subseteq R$, and ask if there is any relation between the initial commuting square and the canonical commuting squares. This need not always be the case, but if we recapture the canonical commuting square, then the commuting square we started with is said to be *flat*.

The first step in investigating flatness or otherwise of a commuting square is the computation of the higher relative commutants; the answer to this computation lies in the following result due to Ocneanu.

THEOREM 5.7.1 (Ocneanu Compactness) *Let A_n, B_n, R_0, R be as above. Then*

$$R_0' \cap R = A_1' \cap B_0. \tag{5.7.3}$$

Proof: First observe, in view of the equations $A_{n+1} = \langle A_n, e_{n+1}\rangle$ and $\{e_{n+1}\}' \cap B_n = B_{n-1}$, that if $m > n \geq 1$, then

$$\begin{aligned} A_m' \cap B_n &= A_n' \cap \{e_{n+1}, e_{n+2}, \cdots, e_m\}' \cap B_n \\ &= A_n' \cap B_{n-1}, \end{aligned}$$

and we thus find that

$$A_m' \cap B_n = A_1' \cap B_0 \quad \forall m > n \geq 1.$$

Since $\bigcup_n A_n$ is weakly dense in R_0, it follows that

$$A_1' \cap B_0 \subseteq R_0' \cap R.$$

(In the following, we identify $a \in R$ with $a\Omega_R$, and thus view R as being metrised by $||\cdot||_2$, the metric in $L^2(R_0)$.)

Let $\mathcal{H}_n$ denote $A_{2n}' \cap B_{2n+1}$, viewed as a (finite-dimensional) Hilbert space (with respect to $||\cdot||_2$), and let E_n (resp., F_n) denote the subspace $A_{2n}' \cap B_{2n}$ (resp., $A_{2n+1}' \cap B_{2n+1}$). Then, as in the proof of Corollary 5.5.5, we can find an (algebra) isomorphism $L_n : \mathcal{H}_0 \to \mathcal{H}_n$ such that $L_n(E_0) = E_n$ and $L_n(F_0) = F_n$.

We first establish the following:

Assertion: There exists a constant $c > 0$ such that

$$c^{-1}||x|| \leq ||L_n(x)|| \leq c||x|| \quad \forall x \in \mathcal{H}_0, \quad \forall n.$$

Proof of assertion: Since L_n is a *-algebra homomorphism, we have to show that the set

$$\{\frac{\operatorname{tr} L_n(x^*x)}{\operatorname{tr} x^*x} : x \in A_0' \cap B_1, n \geq 1\}$$

is bounded away from 0 and ∞. Since $A_0' \cap B_1$ is a finite-dimensional C^*-algebra, it is sufficient to show that for all minimal central projections $p \in A_0, q \in B_1$ (thus the typical minimal central projection in $A_0' \cap B_1$ is pq), and for any one minimal projection $f \in (A_0' \cap B_1)pq$, the sequence $\{\frac{\operatorname{tr} L_n(f)}{\operatorname{tr} f} : n \geq 1\}$ is bounded away from 0 and ∞.

In the following paragraphs, we shall use the notation and terminology of §5.5; in addition, we shall use the following notation: we write $\overline{n} : \pi(A_0) \to \mathbb{N}$ for the dimension-vector for A_0 defined by $\overline{n}_p^2 = \dim A_0p$; further, for $C = A_n$ or B_n, we shall write $\overline{t}^C : \pi(C) \to [0,1]$ for the trace-vector describing $\operatorname{tr}|_C$; thus, it follows from Corollary 5.3.4 that $\overline{t}^{B_{n+2}} = \lambda^{-1}\overline{t}^{B_n}$ and $\overline{t}^{A_{n+2}} = \lambda^{-1}\overline{t}^{A_n}$, where $\lambda = ||G||^2 = ||H||^2$. Finally we shall write $v = \frac{\bar{t}^{A_0}}{||\bar{t}^{A_0}||_2}$, so that v is *the* normalised Perron–Frobenius eigenvector for (GG') corresponding to the eigenvalue λ. In the sequel, we shall need the following standard fact from the Perron-Frobenius theory: under the standing assumption of inclusions being connected, it is true that for any vector w of the same size as v, the sequence $\{(\frac{GG'}{\lambda})^n w\}$ converges to the vector $\langle w, v\rangle v$.

By the construction of L_n, for each n, there exists a system $\{p^{(n)}_{\kappa\circ\beta,\kappa'\circ\beta'} : \kappa\circ\beta, \kappa'\circ\beta' \in \Omega_{(A_0;B_0;B_1)}, f(\beta) = f(\beta'), s(\kappa) = s(\kappa'), f(\kappa) = s(\beta), f(\kappa') = s(\beta')\}$ of matrix units for $A'_{2n}\cap B_{2n+1}$ such that $L_n(p^{(0)}_{\kappa\circ\beta,\kappa'\circ\beta'}) = p^{(n)}_{\kappa\circ\beta,\kappa'\circ\beta'}$. So we have to show that for each fixed path $\kappa\circ\beta \in \Omega_{(A_0;B_0;B_1)}$, the sequence $\{\operatorname{tr} p^{(n)}_{\kappa\circ\beta,\kappa\circ\beta}\}$ is bounded away from 0 and ∞. But we have

$$\begin{aligned}
\operatorname{tr} p^{(n)}_{\kappa\circ\beta,\kappa\circ\beta} &= \sum_{\substack{\theta\in\Omega_{(\mathbb{C};A_{2n})}\\ f(\theta)=s(\kappa)}} \bar{t}^{B_{2n+1}}_{f(\beta)} \\
&= \sum_{p\in\pi(A_0)} \bar{n}_p (\frac{GG'}{\lambda})^n (p, s(\kappa)) \bar{t}^{B_1}_{f(\beta)} \\
&= ((\frac{GG'}{\lambda})^n \bar{n})_{s(\kappa)} \bar{t}^{B_1}_{f(\beta)} \\
&\to \langle \bar{n}, v\rangle v_{s(\kappa)} \bar{t}^{B_1}_{f(\beta)} \\
&> 0,
\end{aligned}$$

thereby proving the assertion.

Before proceeding further, notice that if E, F are subspaces of a finite-dimensional Hilbert space $\mathcal{H}$, the two expressions $p_1(x) = d(x, E\cap F)$ and $p_2(x) = d(x, E) + d(x, F)$ define norms on $\mathcal{H}/(E\cap F)$ and consequently there exists a constant $k > 0$ such that $k^{-1}p_2(x) \le p_1(x) \le kp_2(x)\ \ \forall x \in \mathcal{H}$. In particular, there exists a constant $k_0 > 0$ such that

$$k_0^{-1}(d(x, E_0) + d(x, F_0)) \le d(x, E_0\cap F_0) \le k_0(d(x, E_0) + d(x, F_0))\ \ \forall x \in \mathcal{H}_0. \tag{5.7.4}$$

Coming back to the proof of the theorem, notice first that (as already shown) $E_n \cap F_n = A'_{2n+1}\cap B_{2n} = A'_1 \cap B_0\ \ \forall n$. Suppose now that $x \in R'_0 \cap R$; set $x_n = E_{B_n}x$ for all n; it follows then that $x_n \in A'_n \cap B_n$; i.e., $x_{2n} \in E_n$ and $x_{2n+1} \in F_n$. It follows now from the above, the assertion and (5.7.4) that

$$\begin{aligned}
d(x_{2n}, A'_1\cap B_0) &= d(x_{2n}, E_n\cap F_n) \\
&\le cd(L_n^{-1}(x_{2n}), E_0\cap F_0) \\
&\le ck_0(d(L_n^{-1}(x_{2n}), E_0) + d(L_n^{-1}(x_{2n}), F_0)) \\
&\le c^2k_0(d(x_{2n}, E_n) + d(x_{2n}, F_n)) \\
&= c^2k_0 d(x_{2n}, F_n) \\
&\le c^2k_0||x_{2n} - x_{2n+1}|| \\
&\to 0 \text{ as } n\to\infty.
\end{aligned}$$

On the other hand, since $x_n \to x$ (in $||\cdot||_2$) as $n \to \infty$, this means that $x \in A_1' \cap B_0$, and the proof is complete. □

REMARK 5.7.2 *Theorem 5.7.1 is false if we drop the requirement that the commuting square (5.7.1) is a symmetric commuting square, as shown by the following example: let* $A_1 = A_0 = \mathbb{C} \neq B_0 \neq B_1$; *then (5.7.1) is trivially a commuting square,* $R_0 = \{e_n : n \geq 1\}''$, *and it is not always the case that* $R_0' \cap R = (\{e_1\}' \cap B_0 =)B_0$. *(See the construction of the* $3+\sqrt{3}$ *subfactor in [GHJ].)*

The use of Ocneanu's compactness result lies in the fact that it can be used just as well to compute all the higher relative commutants. In order to see how that goes, we start with a lemma.

LEMMA 5.7.3 *Suppose (5.7.1) is a symmetric commuting square (with respect to the Markov trace). Then there exist a finite set* I, *and* $\{\lambda_i : i \in I\} \subset B_0, \{f_i : i \in I\} \subset A_0$ *such that*

(i) each f_i *is a projection;*

(ii) $E_{A_0}(\lambda_i \lambda_j^*) = \delta_{ij} f_i$;

(iii) $\sum_{i \in I} \operatorname{tr} f_i = ||K||^2$; *and*

(iv) $x = \sum_{i \in I} E_{A_1}(x\lambda_j^*)\lambda_j, \quad \forall x \in B_1$.

Proof: As before, we shall use the notation $\pi(C)$ for the set of minimal central projections of a finite-dimensional C^*-algebra C; for $C \in \{A_0, A_1, B_0, B_1\}$, we write $\overline{t}^C : \pi(C) \to [0,1]$ for the trace-vector corresponding to $\operatorname{tr}|_C$; we shall also write $\overline{n} : \pi(A_0) \to \mathbb{N}$ for the 'dimension-vector' of A_0 defined by $\overline{n}_p^2 = \dim A_0 p$. Thus, for instance, we have

$$KK'\overline{t}^{A_0} = ||K||^2 \overline{t}^{A_0}, \quad \sum_{p \in \pi(A_0)} \overline{n}_p \overline{t}_p^{A_0} = 1. \tag{5.7.5}$$

Finally, we shall denote the typical paths in $\Omega_{(\mathbb{C}:A_0)}, \Omega_{(A_0;B_0)}$ and $\Omega_{(\mathbb{C};A_0;B_0)}$ by the symbols α, κ and β respectively.

Let $\{q_{\beta,\beta'}\}$ denote a system of matrix units for B_0 compatible with the tower $\mathbb{C} \subseteq A_0 \subseteq B_0$. (The following notation is motivated by the fact that the matrix with all entries equal to 1 is customarily denoted by the symbol J.) For each $p \in \pi(A_0)$, define

$$\begin{aligned} j_p &= \frac{1}{\overline{n}_p} \sum_{\substack{\alpha,\alpha' \\ f(\alpha)=f(\alpha')=p}} q_{\alpha,\alpha'} \\ &= \frac{1}{\overline{n}_p} \sum_{\substack{\alpha,\alpha' \\ f(\alpha)=f(\alpha')=p}} \sum_{\kappa, f(\alpha)=s(\kappa)} q_{\alpha\circ\kappa,\alpha'\circ\kappa}. \end{aligned} \tag{5.7.6}$$

(The normalising constant ensures that $j_p \in \mathcal{P}(A_0)$.)

Define $I = \{(\kappa,\beta) : f(\kappa) = f(\beta)\}$ and

$$a_{\kappa,\beta} = \sum_{\alpha, f(\alpha)=s(\kappa)} q_{\alpha o \kappa,\beta} \;\; \forall (\kappa,\beta) \in I. \tag{5.7.7}$$

It follows directly from Proposition 5.4.3 that

$$E_{A_0}(a_{\kappa,\beta} a^*_{\kappa',\beta'}) = \delta_{(\kappa,\beta),(\kappa',\beta')} \frac{\bar{t}^{B_0}_{f(\kappa)}}{\bar{t}^{A_0}_{s(\kappa)}} \bar{n}_{s(\kappa)} j_{s(\kappa)}.$$

Define

$$\begin{aligned} \lambda_{\kappa,\beta} &= \left(\frac{\bar{t}^{B_0}_{f(\kappa)}}{\bar{t}^{A_0}_{s(\kappa)}} \bar{n}_{s(\kappa)} \right)^{-\frac{1}{2}} a_{\kappa,\beta}, \\ f_{\kappa,\beta} &= j_{s(\kappa)}; \;\; \forall (\kappa,\beta) \in I. \end{aligned}$$

It is then clear that $\{\lambda_i, f_i : i \in I\}$ satisfies (i), while (ii) is an immediate consequence of (5.7.5). Further, it follows easily from the definitions that

$$\lambda_i = f_i \lambda_i \;\; \forall i \in I. \tag{5.7.8}$$

Another fairly straightforward computation shows that

$$x = \sum_{i \in I} E_{A_0}(x \lambda_i^*) \lambda_i \;\; \forall x \in B_0. \tag{5.7.9}$$

Now, it follows from Corollary 5.3.4(b)(iv) that B_1 is linearly spanned by $A_1 B_0$; on the other hand, it is a consequence of (5.7.9) that B_0 is linearly spanned by $\bigcup_{i \in I} A_0 \lambda_i$; thus B_1 is linearly spanned by $\bigcup_{i \in I} A_1 \lambda_i$; also, the commuting square condition implies that

$$E_{A_1}(\lambda_i \lambda_j^*) = \delta_{ij} f_i \;\; \forall i, j \in I. \tag{5.7.10}$$

It follows therefore that $\{A_1 \lambda_i : i \in I\}$ is a pairwise orthogonal (with respect to the trace-inner-product) collection of subspaces of B_1 which span B_1. Hence, if $x \in B_1$, there exist $a_i \in A_1, i \in I$, such that $x = \sum_{i \in I} a_i \lambda_i$; it follows from (5.7.10) that $E_{A_1}(x \lambda_j^*) = a_j f_j$, and hence, in view of (5.7.8),

$$\sum_{i \in I} E_{A_1}(x \lambda_i^*) \lambda_i = \sum_{i \in I} a_i f_i \lambda_i = x. \qquad \square$$

COROLLARY 5.7.4 *If $R_0 \subseteq R$ is constructed as in Theorem 5.7.3, and if $\{\lambda_i : i \in I\}$ is as in Lemma 5.7.3, then*

(i) $\{\lambda_i : i \in I\}$ is a basis for R/R_0 as in §4.3; and consequently,

(ii) $[R : R_0] = ||K||^2 = ||L||^2$.

Proof: Since (5.7.2) is a symmetric commuting square for each n, it follows (from the same reasoning employed in the proof of Lemma 5.7.3(iv) and induction) that for any n and for any $x \in B_n$

$$x = \sum_{i \in I} E_{A_n}(x\lambda_i^*)\lambda_i.$$

Since

$$\begin{array}{ccc} B_n & \subset & R \\ \cup & & \cup \\ A_n & \subset & R_0 \end{array}$$

is clearly a commuting square for each n, this means that

$$x = \sum_{i \in I} E_{R_0}(x\lambda_i^*)\lambda_i, \quad \forall x \in \bigcup_{n=1}^{\infty} B_n. \tag{5.7.11}$$

Since both sides of this equation vary continuosly with x, we see that

$$x = \sum_{i \in I} E_{R_0}(x\lambda_i^*)\lambda_i, \quad \forall x \in R,$$

thus proving (i); assertion (ii) follows at once from Lemma 5.7.3(iii). □

Before we start to interpret Ocneanu's compactness theorem to compute the higher relative commutants (and consequently the principal graph) of the subfactor $R_0 \subseteq R$ constructed as above (starting from an arbitrary symmetric commuting square), it will be prudent to adopt a slight change in notation.

We shall henceforth use the symbols A_k^0 and A_k^1, respectively, to denote what we have so far been calling A_k and B_k; likewise, we shall write R_1 for what was earlier called R; thus $R_n = (\bigcup_k A_k^n)'', n = 0, 1$.

Let $R_0 \subseteq R_1 \subseteq R_2 \subseteq \cdots \subseteq R_n \subseteq \cdots$ denote the tower of the basic construction. Since the symbol e_{k+1} has already been reserved for the projection in A_{k+1}^n which implements the conditional expectation of A_k^n onto A_{k-1}^n, for $n = 0, 1$, we shall use the symbol f_{n+1} to denote the projection in R_{n+1} which implements the conditional expectation of R_n onto R_{n-1}, for $n \geq 1$. (Again, the sequence starts only with f_2.)

For $n \geq 1$, inductively define

$$A_k^{n+1} = \langle A_k^n, f_{n+1} \rangle, \quad \forall k \geq 0.$$

PROPOSITION 5.7.5 *Consider the grid $\{A_k^n : n, k \geq 0\}$ constructed as above. Let $n \geq 0$.*

$(i)_n$ A_k^n is a finite-dimensional C^*-algebra, for each $k \geq 0$, and $R_n = (\bigcup_{k=1}^{\infty} A_k^n)''$.

$(ii)_n$ If $n \geq 1$, then for each $k \geq 0$,

$$\begin{array}{ccc} A_k^n & \subset & A_{k+1}^n \\ \cup & & \cup \\ A_k^{n-1} & \subset & A_{k+1}^{n-1} \end{array} \tag{5.7.12}$$

is a symmetric commuting square with respect to $\mathrm{tr}_{R_n}|_{A^n_{k+1}}$ *(which is also the Markov trace).*

$(iii)_n$ *If* $n \geq 2$, *then, for each* $k \geq 0$,

$$A^{n-2}_k \subseteq A^{n-1}_k \subseteq A^n_k$$

is an instance of the basic construction, and the projection f_n *implements the conditional expectation of* A^{n-1}_k *onto* A^{n-2}_k.

$(iv)_n$ *For each* $k \geq 1$,

$$A^n_{k-1} \subseteq A^n_k \subseteq A^n_{k+1}$$

is an instance of the basic construction, and the projection e_{k+1} *implements the conditional expectation of* A^n_k *onto* A^n_{k-1}.

Proof: We prove the proposition by induction on n. The proposition being true, by construction, for $n = 0, 1$, we assume that $n \geq 2$ and that the assertions $(\mathrm{i})_{n-1}$–$(\mathrm{iv})_{n-1}$ are valid.

It follows from $(\mathrm{ii})_{n-1}$ that

$$\begin{array}{ccc} A^{n-1}_k & \subset & R_{n-1} \\ \cup & & \cup \\ A^{n-2}_k & \subset & R_{n-2} \end{array} \tag{5.7.13}$$

is a commuting square, for each k; hence, by the defining property of f_n, we find that if $a^{n-1}_k \in A^{n-1}_k$, then

$$f_n a^{n-1}_k f_n = E_{R_{n-2}}(a^{n-1}_k) f_n = E_{A^{n-2}_k}(a^{n-1}_k) f_n$$

and hence f_n indeed implements the conditional expectation of A^{n-1}_k onto A^{n-2}_k; in particular, it follows that the set $\{a_0 + \sum_{i=1}^m a_i f_n a'_i : m \in \mathbb{N}, a_i, a'_i \in A^{n-1}_k\}$ is a finite-dimensional C^*-algebra containing $(A^{n-1}_k \cup \{f_n\})$, and consequently

$$A^n_k = \{a_0 + \sum_{i=1}^m a_i f_n a'_i : m \in \mathbb{N}, a_i, a'_i \in A^{n-1}_k\}, \tag{5.7.14}$$

thus establishing $(\mathrm{i})_n$.

Since $E_{A^{n-1}_k}(f_n) = E_{A^{n-1}_k}(E_{R_{n-1}}(f_n)) = \tau$, where $\tau = [R_1 : R_0]^{-1}$, the equation (5.7.14) implies that $E_{A^{n-1}_{k+1}}(A^n_k) \subseteq A^{n-1}_k$, and consequently (5.7.12) is indeed a commuting square with respect to $\mathrm{tr}_{R_n}|_{A^n_{k+1}}$. In particular, since this is valid for all k, it is also true that

$$\begin{array}{ccc} A^n_k & \subset & R_n \\ \cup & & \cup \\ A^{n-1}_k & \subset & R_{n-1} \end{array}$$

is a commuting square with respect to tr_{R_n}.

Let $\{\lambda_i : i \in I\} \subset A_0^1$ be a basis for R_1/R_0 as in Corollary 5.7.4(i). It follows from Lemma 4.3.4(ii) that $\{\tau^{-\frac{n-1}{2}} f_n f_{n-1} \cdots f_2 \lambda_i : i \in I\}$ is a basis for R_n/R_{n-1}. In particular, if $k \geq 1$, and $a_k^n \in A_k^n$, then

$$\begin{aligned} a_k^n &= \tau^{-(n-1)} \sum_{i \in I} E_{R_{n-1}}(a_k^n \lambda_i^* f_2 \cdots f_n) f_n \cdots f_2 \lambda_i \\ &= \tau^{-(n-1)} \sum_{i \in I} E_{A_k^{n-1}}(a_k^n \lambda_i^* f_2 \cdots f_n) f_n \cdots f_2 \lambda_i. \end{aligned} \tag{5.7.15}$$

Hence

$$A_k^n = \bigvee A_k^{n-1} \cdot A_{k-1}^n, \tag{5.7.16}$$

where the symbol $\bigvee$ denotes 'span'. It follows from Remark 5.3.5 that $z_{A_{k-1}^n}(f_n) = 1$, and hence, by Corollary 5.3.2, we have proved $(\text{iii})_n$.

Notice next that $k \geq 1 \Rightarrow f_n \cdots f_2 \lambda_i \in A_0^n \subseteq A_{k-1}^n$; so, in view of the already established fact that (5.7.12) is a commuting square, we see from (5.7.15) that for arbitrary $a_k^n \in A_k^n$,

$$E_{A_{k-1}^n}(a_k^n) = \tau^{-(n-1)} \sum_{i \in I} E_{A_{k-1}^{n-1}}(a_k^n \lambda_i^* f_2 \cdots f_n) f_n \cdots f_2 \lambda_i;$$

on the other hand, since $f_n \cdots f_2 \lambda_i \in A_0^n \subseteq \{e_k\}'$, it follows from $(\text{iv})_{n-1}$ that

$$\begin{aligned} e_{k+1} a_k^n e_{k+1} &= \tau^{-(n-1)} \sum_{i \in I} E_{A_{k-1}^{n-1}}(a_k^n \lambda_i^* f_2 \cdots f_n) f_n \cdots f_2 \lambda_i e_{k+1} \\ &= E_{A_{k-1}^n}(a_k^n) e_{k+1}; \end{aligned} \tag{5.7.17}$$

hence e_{k+1} indeed implements the conditional expectation of A_k^n onto A_{k-1}^n. Deduce next from equation (5.7.16) and $(\text{iv})_{n-1}$ that

$$\begin{aligned} A_{k+1}^n &= \bigvee A_{k+1}^{n-1} A_k^n \\ &= \bigvee A_k^{n-1} e_{k+1} A_k^{n-1} A_k^n \\ &= \bigvee A_k^{n-1} e_{k+1} A_k^n \\ &\subseteq \bigvee A_k^n e_{k+1} A_k^n, \end{aligned}$$

which implies that $z_{A_{k+1}^n}(e_{k+1}) = 1$, which establishes $(\text{iv})_n$ in view of Corollary 5.3.2.

Finally, in view of the already establlished fact that (5.7.12) is a commuting square for all k, we see from the parenthetical remark in $(\text{ii})_{n-1}$, and from $(\text{iv})_{n-1}$, that

$$E_{A_k^n}(e_{k+1}) = E_{A_k^{n-1}}(e_{k+1}) \in \mathbb{C}1.$$

In view of the already established $(\text{iv})_n$, this means that $\text{tr}_{R_n}|_{A_k^n}$ is a Markov trace for the inclusion $A_{k-1}^n \subseteq A_k^n$. The already established fact that (5.7.12) is a commuting square, the identity (5.7.16), and the previous sentence complete the proof of $(\text{ii})_n$, and with that, the induction step, and consequently the proof of the proposition, is complete. □

We are finally ready to use Ocneanu's compactness theorem to compute the higher relative commutants.

THEOREM 5.7.6 (Ocneanu compactness (*contd.*))
Let

$$\begin{array}{ccccccccccccc}
\cdots & \subseteq & \cdots & \subseteq & \cdots & \subseteq & \cdots & \subseteq & \cdots & \subseteq & \cdots & \to & \cdots \\
\cup & & \cup & & \cup & & \cup & & \cup & & & & \cup \\
A_0^n & \subseteq & A_1^n & \subseteq & A_2^n & \subseteq & \cdots & \subseteq & A_k^n & \subseteq & \cdots & \to & R_n \\
\cup & & \cup & & \cup & & \cup & & \cup & & & & \cup \\
\cdots & \subseteq & \cdots & \subseteq & \cdots & \subseteq & \cdots & \subseteq & \cdots & \subseteq & \cdots & \to & \cdots \\
\cup & & \cup & & \cup & & \cup & & \cup & & & & \cup \\
A_0^1 & \subseteq & A_1^1 & \subseteq & A_2^1 & \subseteq & \cdots & \subseteq & A_k^1 & \subseteq & \cdots & \to & R_1 \\
\cup & & \cup & & \cup & & \cup & & \cup & & & & \cup \\
A_0^0 & \subseteq & A_1^0 & \subseteq & A_2^0 & \subseteq & \cdots & \subseteq & A_k^0 & \subseteq & \cdots & \to & R_0
\end{array}$$

be as in Proposition 5.7.5. Then, for each $n \geq 1$,

$$R_0' \cap R_n = A_1^{0\prime} \cap A_0^n.$$

Proof: Fix $n \geq 1$ and consider the grid

$$\begin{array}{ccccccccccccc}
A_0^n & \subseteq & A_1^n & \subseteq & A_2^n & \subseteq & \cdots & \subseteq & A_k^n & \subseteq & \cdots & \to & R_n \\
\cup & & \cup & & \cup & & \cup & & \cup & & & & \cup \\
A_0^0 & \subseteq & A_1^0 & \subseteq & A_2^0 & \subseteq & \cdots & \subseteq & A_k^0 & \subseteq & \cdots & \to & R_0 .
\end{array}$$

By Proposition 5.7.5, this grid satisfies the hypothesis of Theorem 5.7.3, and hence, by that theorem, the desired conclusion follows. □

We now state a few related problems whose solutions we would like to see.

QUESTION 5.7.7 *Given a symmetric commuting square, say (5.7.1), does there exist an algorithm which, in the preceding notation, would:*

(a) compute $\dim_{\mathbb{C}}(R_0' \cap R_n)$ *in polynomial time (as a function of the input data)?*

(b) decide in finite time if the subfactor $R_0 \subseteq R_1$ *has finite depth?*

(c) answer (b) in polynomial time?

We should mention here that Ocneanu has shown that if it is already known that the depth is a finite integer, say n, then the dimensions of the higher relative commutants can be computed in polynomial time.

Chapter 6

Vertex and spin models

6.1 Computing higher relative commutants

This chapter is devoted to a discussion of the subfactors arising, as in §4.4, from an initial commuting square which is a 'vertex model' in the terminology of §5.2.3. Actually, we shall work with a slight generalisation, as follows.

LEMMA 6.1.1 *Let*

$$\begin{array}{ccc} B_0 & \overset{[N]}{\subset} & B_1 \\ {}_{[k]}\cup & & \cup_{[k]} \\ A_0 & \overset{[N]}{\subset} & A_1 \end{array} \tag{6.1.1}$$

be a (clearly symmetric) commuting square of finite-dimensional C^*-algebras such that the inclusion matrices are 1×1 matrices (as indicated) and where $A_0 = \mathbb{C}$. Suppose this commuting square is described by the biunitary matrix u (see §5.5), which is a $kN \times Nk$ matrix given by

$$u = ((u_{\alpha a}^{b\beta})), 1 \leq \alpha, \beta \leq N, 1 \leq a, b \leq k.$$

Define the unitary matrix $w \in M_{Nk}(\mathbb{C})$ by

$$w_{\beta b}^{\alpha a} = \overline{u_{\alpha a}^{b\beta}}. \tag{6.1.2}$$

Then the square (6.1.1) is isomorphic to the square

$$\begin{array}{ccc} w(1 \otimes M_k(\mathbb{C}))w^* & \subset & M_N(\mathbb{C}) \otimes M_k(\mathbb{C}) \\ \cup & & \cup \\ \mathbb{C} & \subset & M_N(\mathbb{C}) \otimes 1. \end{array} \tag{6.1.3}$$

Proof: Let $\{p_{\alpha a,\alpha' a'} : 1 \leq \alpha, \alpha' \leq N, 1 \leq a, a' \leq k\}$ and $\{q_{b\beta,b'\beta'} : 1 \leq \beta, \beta' \leq N, 1 \leq b, b' \leq k\}$ be systems of matrix units for B_1 compatible with the towers $A_0 \subset A_1 \subset B_1$ and $A_0 \subset B_0 \subset B_1$ respectively, with respect to which the biunitary matrix u describes the given commuting square.

The system $\{p_{\alpha a,\alpha' a'} : 1 \leq \alpha, \alpha' \leq N, 1 \leq a, a' \leq k\}$ yields an isomorphism $\psi : B_1 \to M_N(\mathbb{C}) \otimes M_k(\mathbb{C})$ thus: $\psi(p_{\alpha a,\alpha' a'}) = e_{\alpha\alpha'} \otimes e_{aa'}$, where the e's denote the standard systems of matrix units in matrix algebras. By definition, the algebra A_1 has a system of matrix units defined by $p_{\alpha\alpha'} = \sum_{a=1}^{k} p_{\alpha a,\alpha' a}$; hence $\psi(p_{\alpha\alpha'}) = e_{\alpha\alpha'} \otimes 1$; i.e., $\psi(A_1) = M_N(\mathbb{C}) \otimes 1$.

By definition, we have

$$p_{\alpha a,\alpha' a'} = \sum_{\beta,\beta'=1}^{N} \sum_{b,b'=1}^{k} u_{\alpha a}^{b\beta} \overline{u}_{\alpha' a'}^{b'\beta'} q_{b\beta,b'\beta'};$$

and hence

$$q_{b\beta,b'\beta'} = \sum_{\alpha,\alpha'=1}^{N} \sum_{a,a'=1}^{k} w_{\beta b}^{\alpha a} \overline{w}_{\beta' b'}^{\alpha' a'} p_{\alpha a,\alpha' a'},$$

and consequently,

$$\begin{aligned} \psi(q_{b\beta,b'\beta'}) &= \sum_{\alpha,\alpha'=1}^{N} \sum_{a,a'=1}^{k} w_{\beta b}^{\alpha a} \overline{w}_{\beta' b'}^{\alpha' a'} e_{\alpha\alpha'} \otimes e_{aa'} \\ &= w(e_{\beta\beta'} \otimes e_{bb'})w^*, \end{aligned}$$

where we view w as an element of $M_N(\mathbb{C}) \otimes M_k(\mathbb{C})$ in the obvious manner. But B_0 is spanned by $\{q_{bb'} : 1 \leq b, b' \leq k\}$, where $q_{bb'} = \sum_{\beta=1}^{N} q_{b\beta,b'\beta}$. It follows at once that

$$\psi(B_0) = w(1 \otimes M_k(\mathbb{C}))w^*$$

as desired. □

DEFINITION 6.1.2 *We shall refer to any commuting square of the form (6.1.1) as a vertex model.*

REMARK 6.1.3 *We shall need the fact, which the proof of the lemma shows, that conversely if $w = ((w_{\beta b}^{\alpha a})) \in M_N(\mathbb{C}) \otimes M_k(\mathbb{C})$ is unitary, and if*

$$\begin{array}{ccc} M_N(\mathbb{C}) \otimes 1 & \subseteq & M_N(\mathbb{C}) \otimes M_k(\mathbb{C}) \\ \cup & & \cup \\ \mathbb{C} & \subseteq & w(1 \otimes M_k(\mathbb{C}))w^* \end{array} \qquad (6.1.4)$$

is a commuting square (and is hence a vertex model in our terminology), then the biunitary matrix u which describes this commuting square is given by $u = ((u_{b\beta}^{\alpha a}))$, where $u_{b\beta}^{\alpha a} = w_{\beta b}^{\alpha a}$.

In the rest of this section, we shall start with a vertex model, as given by (6.1.3), think of this as the initial commuting square

$$\begin{array}{ccc} A_0^1 & \subseteq & A_1^1 \\ \cup & & \cup \\ A_0^0 & \subseteq & A_1^0, \end{array} \qquad (6.1.5)$$

and describe the tower $\{R_0' \cap R_n : n \geq 1\}$ of relative commutants, as discussed in §4.4. We shall use the notation of that section; in addition, we shall find it convenient to use the notation

$$C_n = R_0' \cap R_n, n \geq 1.$$

By Theorem 5.7.6, we have

$$C_n = A_1^{0\prime} \cap A_0^n.$$

On the other hand, the grid $\{A_k^n : n \geq 0, k = 0, 1\}$ (when flipped over by 90° from the way it looks in the grid in the statement of Theorem 5.7.6) is given by

$$\begin{array}{ccccccccccc} A_1^0 & \subseteq & A_1^1 & \subseteq & \langle A_1^1, f_2\rangle & \subseteq & \langle A_1^1, f_2, f_3\rangle & \subseteq & \cdots & \subseteq & \cdots \\ \cup & & \cup & & \cup & & \cup & & & & \\ A_0^0 & \subseteq & A_0^1 & \subseteq & \langle A_0^1, f_2\rangle & \subseteq & \langle A_0^1, f_2, f_3\rangle & \subseteq & \cdots & \subseteq & \cdots; \end{array} \tag{6.1.6}$$

the latter is, by Proposition 5.7.5(iii), the result of repeated applications of the basic construction to the top row of the vertex model given by (6.1.4).

According to Remark 6.1.3, the basic (initial) commuting square

$$\left(\begin{array}{ccc} A_1^0 & \subseteq & A_1^1 \\ \cup & & \cup \\ A_0^0 & \subseteq & A_0^1 \end{array}\right) = \left(\begin{array}{ccc} M_N(\mathbb{C}) \otimes 1 & \subseteq & M_N(\mathbb{C}) \otimes M_k(\mathbb{C}) \\ \cup & & \cup \\ \mathbb{C} & \subseteq & w(1 \otimes M_k(\mathbb{C}))w^* \end{array}\right)$$

of the grid (6.1.6) is described by the $Nk \times kN$ biunitary matrix $u = ((u_{b\beta}^{\alpha a}))$, where $u_{b\beta}^{\alpha a} = w_{\beta b}^{\alpha a}$.

It follows from Corollary 5.5.5(i) and Proposition 5.5.4(b) that, for each $n \geq 1$, the commuting square

$$\begin{array}{ccc} A_1^0 & \overset{[k^n]}{\subseteq} & A_1^n \\ {\scriptstyle [N]}\cup & & \cup{\scriptstyle [N]} \\ A_0^0 & \overset{[k^n]}{\subseteq} & A_0^n \end{array}$$

is described by the $Nk^n \times k^n N$ biunitary matrix $U_{[0,n]}$, which is defined by

$$\begin{aligned} (U_{[0,n]})_{b_1 \cdots b_n \beta}^{\alpha a_1 \cdots a_n} &= \sum_{\gamma_1, \cdots, \gamma_{n-1}=1}^{N} u_{b_1 \gamma_1}^{\alpha a_1} (V(u))_{b_2 \gamma_2}^{\gamma_1 a_2} u_{b_3 \gamma_3}^{\gamma_2 a_3} \cdots (n \text{ factors}) \\ &= \sum_{\gamma_1, \cdots, \gamma_{n-1}=1}^{N} u_{b_1 \gamma_1}^{\alpha a_1} \overline{u_{b_2 \gamma_1}^{\gamma_2 a_2}} u_{b_3 \gamma_3}^{\gamma_2 a_3} \cdots (n \text{ factors}) \\ &= \sum_{\gamma_1, \cdots, \gamma_{n-1}=1}^{N} w_{\gamma_1 b_1}^{\alpha a_1} \overline{w_{\gamma_1 b_2}^{\gamma_2 a_2}} w_{\gamma_3 b_3}^{\gamma_2 a_3} \cdots (n \text{ factors}). \end{aligned}$$

(We pause to point out that in the n-th factors of the product in each of three preceding lines, the subscript is given by $b_n\beta$ in the first two lines and by βb_n in the last line.)

By another application of Remark 6.1.3, we find that we may make the identification

$$\begin{pmatrix} A_1^0 & \subseteq & A_1^n \\ \cup & & \cup \\ A_0^0 & \subseteq & A_0^n \end{pmatrix} = \begin{pmatrix} M_N(\mathbb{C}) \otimes 1 & \subseteq & M_N(\mathbb{C}) \otimes (\otimes^n M_k(\mathbb{C})) \\ \cup & & \cup \\ \mathbb{C} & \subseteq & W_{[0,n]}(1 \otimes (\otimes^n M_k(\mathbb{C})))W_{[0,n]}^* \end{pmatrix}, \tag{6.1.7}$$

where $W_{[0,n]} = (((W_{[0,n]})_{\beta b_1 \cdots b_n}^{\alpha a_1 \cdots a_n})) \in M_{Nk^n}(\mathbb{C}) (= M_N(\mathbb{C}) \otimes (\otimes^n M_k(\mathbb{C})))$ is defined by

$$(W_{[0,n]})_{\beta b_1 \cdots b_n}^{\alpha a_1 \cdots a_n} = (U_{[0,n]})_{b_1 \cdots b_n \beta}^{\alpha a_1 \cdots a_n}.$$

Hence it follows that $C_n = A_1^{0'} \cap A_0^n = (1 \otimes (\otimes^n M_k(\mathbb{C}))) \cap W_{[0,n]}(1 \otimes (\otimes^n M_k(\mathbb{C})))W_{[0,n]}^*$. Thus we see that C_n may be identified with the set of those $F = ((F_{b_1 \cdots b_n}^{a_1 \cdots a_n})) \in \otimes^n M_k(\mathbb{C})$ for which there exists a $G = ((G_{b_1 \cdots b_n}^{a_1 \cdots a_n})) \in \otimes^n M_k(\mathbb{C})$ such that $1 \otimes F = W_{[0,n]}(1 \otimes G)W_{[0,n]}^*$; in longhand, this means that for all $\alpha, \beta \in \{1, 2, \cdots, N\}, a_1, b_1, \cdots, a_n, b_n \in \{1, 2, \cdots, k\}$, we have

$$\begin{aligned} &\sum_{c_1, \cdots, c_n = 1}^{k} \sum_{\gamma_1, \cdots, \gamma_{n-1} = 1}^{N} F_{c_1 \cdots c_n}^{a_1 \cdots a_n} \{ w_{\gamma_1 b_1}^{\alpha c_1} \overline{w_{\gamma_1 b_2}^{\gamma_2 c_2}} w_{\gamma_3 b_3}^{\gamma_2 c_3} \cdots (n \text{ factors}) \} \\ = &\sum_{c_1, \cdots, c_n = 1}^{k} \sum_{\gamma_1, \cdots, \gamma_{n-1} = 1}^{N} \{ w_{\gamma_1 c_1}^{\alpha a_1} \overline{w_{\gamma_1 c_2}^{\gamma_2 a_2}} w_{\gamma_3 c_3}^{\gamma_2 a_3} \cdots (n \text{ factors}) \} G_{b_1 \cdots b_n}^{c_1 \cdots c_n} \end{aligned} \tag{6.1.8}$$

We shall now re-write the answer to the computation of the higher relative commutants – as provided by equation (6.1.8) – in a diagrammatical fashion that should begin to convince the reader that these examples do indeed have a connection with the vertex models of statistical mechanics. (The letter w that appears above will, in the latter picture, correspond to Boltzmann weights.)

We shall be drawing diagrams as in the theory of knots – more precisely, there will be over- and under-crossings, as below:

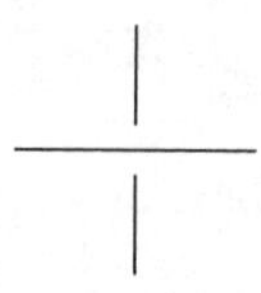

Actually, we shall be dealing with such diagrams where the two strands of the crossing are both oriented, and we shall follow the convention of knot theory and refer to the two different possible situations as *positive* and *negative* crossings. Also, typically, the four ends of the strands will be labelled, with the ends of the over-crossing labelled by Greek alphabets (which are assumed to vary over the set $\{1, 2, \cdots, N\}$), while the ends of the under-crossing will be labelled by Latin alphabets (which are assumed to vary over the set

$\{1, 2, \cdots, k\}$). We assign a 'Boltzmann weight' to such a labelled oriented crossing by the following prescription:

Positive Crossing

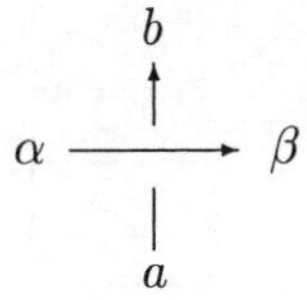

$\mapsto \quad w^{\alpha b}_{\beta a}$

Negative Crossing

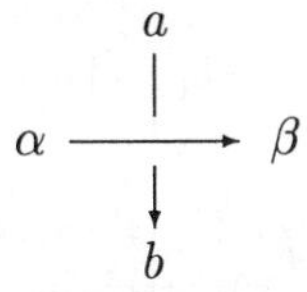

$\mapsto \quad \overline{w^{\beta a}_{\alpha b}}$

(One way to remember the convention is as follows: for either type of crossing, the Greek (resp., Latin) indices of w correspond to the over- (resp., under-) strand of the crossing; for a positive crossing, the orientation of the over- (resp., under-) strand is given by reading the Greek (resp., Latin) indices from top to bottom (resp., bottom to top); for a negative crossing, the complex conjugate appears, and the orientations of the two strands are both read in the opposite way to that of the positive crossing.)

With the above conventions, equation (6.1.8) may be rephrased as the equality

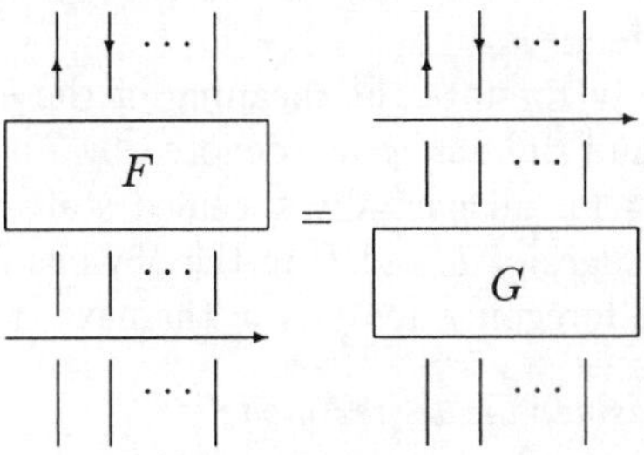

(6.1.9)

where the vertical strings (on each side of the equality) are alternately oriented upwards and downwards (starting with the one at the extreme left), and the diagram is interpreted as follows:

By a state σ of a diagram such as the following – where we have chosen $n = 2$ for simplicity –

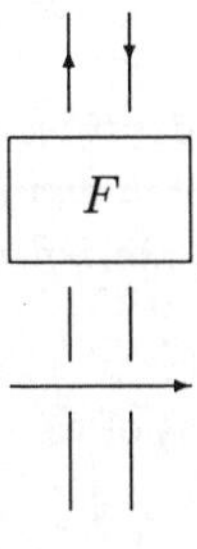

T

– we shall mean a specification of a Latin (resp., Greek) letter to each segment of a vertical (resp., horizontal) string. For example, σ might be specified as follows:

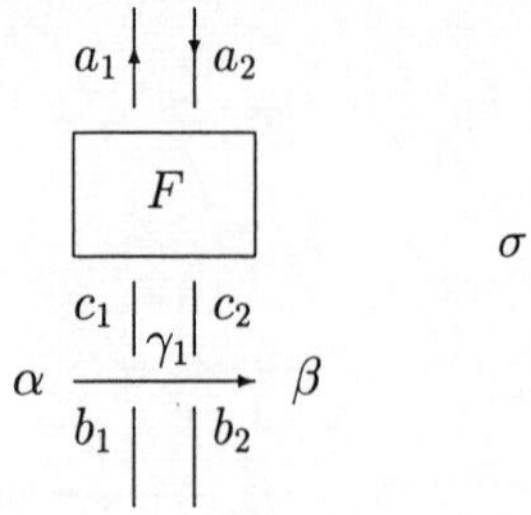

a_1 a_2

F

σ

c_1 c_2

γ_1

α β

b_1 b_2

By definition, the 'energy $E(\sigma)$' associated to such a state σ is the product of the indicated matrix entry with the product of the Boltzmann weights of all the crossings; thus, in the above example, we have

$$E(\sigma) = F^{a_1 a_2}_{c_1 c_2} w^{\alpha c_1}_{\gamma_1 b_1} \overline{w^{\beta c_2}_{\gamma_1 b_2}}.$$

By a boundary edge of a diagram (such as T above), we shall mean any segment of a strand, at least one of whose edges is 'free'; thus, in the preceding example, the edges labelled $a_1, a_2, b_1, b_2, \alpha$ and β are the boundary edges.

By a 'partial state' will be meant a 'state' which has been specified only on the boundary edges. Given a partial state κ on a diagram T, define the 'partition function' Z^κ_T to be the sum of the energies $E(\sigma)$ corresponding to all states σ which 'extend' κ.

We are finally ready to state the meaning of the equality (6.1.9); if the diagrams on the left and right sides are denoted by T and B respectively, we require that $Z^\kappa_T = Z^\kappa_B$ for all partially specified states κ, where we identify the sets of boundary edges of T and B in the obvious fashion.

We summarise the foregoing analysis in the next theorem.

THEOREM 6.1.4 *Consider the vertex model*

$$\begin{pmatrix} A^1_0 & \subseteq & A^1_1 \\ \cup & & \cup \\ A^0_0 & \subseteq & A^0_1 \end{pmatrix} = \begin{pmatrix} w(1 \otimes M_k(\mathbb{C}))w^* & \subseteq & M_N(\mathbb{C}) \otimes M_k(\mathbb{C}) \\ \cup & & \cup \\ \mathbb{C} & \subseteq & M_N(\mathbb{C}) \otimes 1 \end{pmatrix}. \quad (6.1.10)$$

Let $R_0 \subseteq R_1$ be the subfactor constructed from this initial commuting square as in §4.4, and let $R_0 \subseteq R_1 \subseteq R_2 \subseteq \cdots \subseteq R_n \subseteq \cdots$ be the tower of the basic construction.

Then, for $n \geq 1$, $R_0' \cap R_n$ can be identified with the set of those matrices $F = ((F^{a_1 \cdots a_n}_{b_1 \cdots b_n})) \in \bigotimes^n M_k(\mathbb{C})$ for which there exists a $G = ((G^{a_1 \cdots a_n}_{b_1 \cdots b_n})) \in \bigotimes^n M_k(\mathbb{C})$ such that equation (6.1.9) is satisfied (in the sense just discussed). □

We now show how the computations of this section carry over, with slight modifications, to spin models.

Analogous to Lemma 6.1.1, one can prove, quite easily, that if

$$\begin{array}{ccc} B_0 & \overset{G'}{\subset} & B_1 \\ {}_G\cup & & \cup_{G'} \\ A_0 & \overset{G}{\subset} & A_1 \end{array} \tag{6.1.11}$$

is a (clearly symmetric) commuting square with inclusions as indicated, where $G = [1\ 1 \cdots 1]$ is the $1 \times N$ matrix with all entries equal to 1, if this commuting square is described by the biunitary matrix $u = ((u^i_j))$, and if $A_0 = \mathbb{C}$, then (the matrix u is a complex Hadamard matrix – see §5.2.2 – and) the square (6.1.11) is isomorphic to

$$\begin{array}{ccc} \Delta & \subseteq & M_N(\mathbb{C}) \\ \cup & & \cup \\ \mathbb{C} & \subset & u\Delta u^*, \end{array} \tag{6.1.12}$$

where Δ denotes the diagonal subalgebra of $M_N(\mathbb{C})$. Thus, these are precisely what we called 'spin models' in §5.2.2.

Suppose conversely that we start with a spin model, which we assume for notational reasons is given by

$$\begin{array}{ccc} w\Delta w^* & \subseteq & M_N(\mathbb{C}) \\ \cup & & \cup \\ \mathbb{C} & \subset & \Delta; \end{array} \tag{6.1.13}$$

if we think of this as the initial commuting square (6.1.5), and construct the tower $\{R_n : n \geq 0\}$ as in §4.4, then an analysis akin to the one given above for vertex models is seen to show that the higher relative commutants continue to be described by the validity of equation (6.1.9), only the interpretation of this equality is now as follows: (we give the general description first, then illustrate with the cases of the second and third relative commutants, and the reader should then be convinced that this diagrammatical description is much more compact and clean than an explicit one with mathematical symbols; among other things, such an explicit description would require different descriptions according to the parity of n.)

To start with, we ignore orientations on the vertical strands. Next, given a diagram – such as the left or right side of (6.1.9) – first colour the connected components of the diagram alternately black and white, in a 'chequer-board' fashion, with the convention that the component at the south-west corner is shaded black. By a state σ of such a diagram, we now mean an assignment of a symbol from the set $\{1, 2, \cdots, N\}$ to each of the components coloured black. By a boundary component, we mean one which is unbounded, and by a partially defined state, we mean a state which has been specified only on the boundary components.

Once a state σ has been fixed, we assign Boltzmann weights to each crossing (which must clearly be one of the following types) as follows:

Positive Crossing $\mapsto$ w_b^a

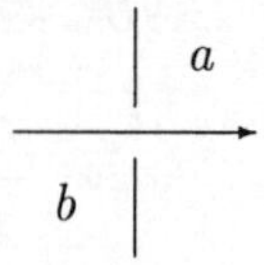

Negative Crossing $\mapsto$ $\overline{w_b^a}$

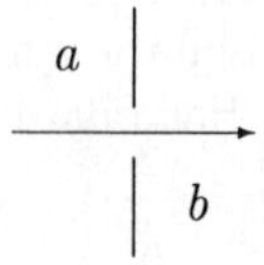

And we define the energy $E(\sigma)$ of the state σ to be the product of the indicated 'matrix entry' with the product of the Boltzmann weights of all the crossings.

Given a partially defined state κ on a diagram T, we define the partition function Z_T^κ to be the sum of the energies $E(\sigma)$ of all states σ which extend κ. Finally, the equality (6.1.9), in the context of spin models, means that once the boundary components of the diagrams T and B of the left and right sides of (6.1.9) have been identified in the obvious fashion, then $Z_T^\kappa = Z_B^\kappa$ for all partially defined states κ.

Since the precise interpretation of this answer depends upon the parity of the relative commutant in question, we illustrate the preceding discussion by explicitly writing out the second and third relative commutants.

$n = 2:$

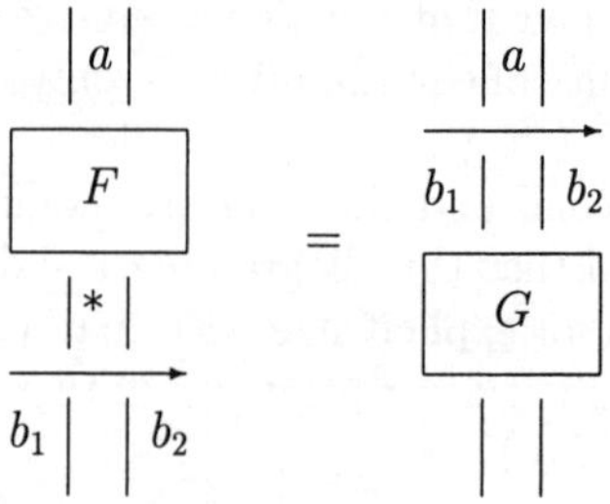

This means that $R_0' \cap R_2$ consists of those $F = ((F_b^a)) \in M_N(\mathbb{C})$ such that there exists an array of numbers $G = ((G_{b_1 b_2}))$ (which should be actually thought of as an element of C^{N^2}) such that for all choices of $a, b_1, b_2 \in \{1, 2, \cdots, N\}$,

$$\sum_{x=1}^{N} F_x^a w_{b_1}^x \overline{w_{b_2}^x} = w_{b_1}^a \overline{w_{b_2}^a} G_{b_1 b_2}.$$

$n = 3$:

a_1 a_2 a_1 a_2

F $=$ $*$

$*$ G

b_1 b_2 b_1 b_2

This means that $R_0' \cap R_3$ consists of those arrays of numbers $F = ((F^{a_1a_2}_{xa_2}))$ (which must be thought of as an element of a direct sum of N copies of $M_N(\mathbb{C})$) such that there exists an array of numbers $G = ((G^{b_1x}_{b_1b_2}))$ (which should also be thought of as an element of a direct sum of N copies of $M_N(\mathbb{C})$), such that, for all choices of $a_1, a_2, b_1, b_2 \in \{1, 2, \cdots, N\}$,

$$\sum_{x=1}^{N} F^{a_1a_2}_{xa_2} w^x_{b_1} \overline{w^x_{b_2}} w^{a_2}_{b_2} = \sum_{x=1}^{N} w^{a_1}_{b_1} \overline{w^{a_1}_x} w^{a_2}_x G^{b_1x}_{b_1b_2}.$$

We summarise the foregoing in the following result.

THEOREM 6.1.5 *Consider the spin model*

$$\begin{pmatrix} A_0^1 & \subseteq & A_1^1 \\ \cup & & \cup \\ A_0^0 & \subseteq & A_1^0 \end{pmatrix} = \begin{pmatrix} w\Delta w^* & \subseteq & M_N(\mathbb{C}) \\ \cup & & \cup \\ \mathbb{C} & \subset & \Delta \end{pmatrix}. \tag{6.1.14}$$

Let $R_0 \subseteq R_1$ be the subfactor constructed from this initial commuting square as in §4.4, and let $R_0 \subseteq R_1 \subseteq R_2 \subseteq \cdots \subseteq R_n \subseteq \cdots$ be the tower of the basic construction.

Then, for $n \geq 1$, $R_0' \cap R_n$ can be identified with the set of those 'matrices' F for which there exists a 'matrix' G such that equation (6.1.9) is satisfied (in the sense just discussed). □

REMARK 6.1.6 *In the diagrammatic framework described above – for both vertex and spin models – it should be remarked that the biunitarity condition on the matrix w is exactly equivalent to the requirement that 'the partition function given by w is invariant with respect to Reidemeister moves of type II' – meaning that, no matter how the two strands are oriented, resp., the regions are shaded black and white, we have*

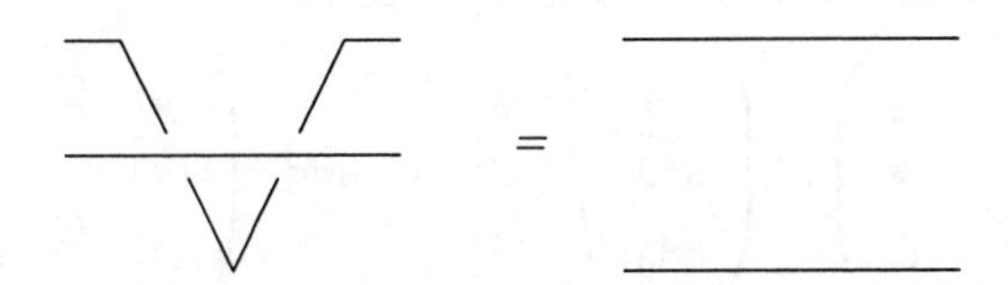

The verification that this is indeed the case will be a good exercise for the reader to ensure that (s)he has understood our procedure.

REMARK 6.1.7 *It is easily verified that, in the case of a vertex model, the embeddings* $(R_0' \cap R_n) \hookrightarrow \bigotimes^n M_k(\mathbb{C})$ *which we obtained are consistent in the sense that the following is a commutative diagram of inclusions (where we embed* $\bigotimes^n M_k(\mathbb{C})$ *into* $\bigotimes^{n+1} M_k(\mathbb{C})$ *by* $x \mapsto x \otimes 1$*):*

$$\begin{array}{ccc} (R_0' \cap R_{n+1}) & \hookrightarrow & \bigotimes^{n+1} M_k(\mathbb{C}) \\ \cup & & \cup \\ (R_0' \cap R_n) & \hookrightarrow & \bigotimes^n M_k(\mathbb{C}). \end{array}$$

Consequently, we may effectively use Theorem 6.1.4 for even computing the principal graph of a subfactor arising from a vertex model.

For identical reasons, an analogous remark is also valid for spin models.

REMARK 6.1.8 *The purpose of this remark is to point out that if* $R_0 \subseteq R_1$ *is constructed as above from a vertex (resp., spin) model, then so also is* $R_1 \subseteq R_2$*, and to consequently derive some facts concerning the dual principal graph.*

(i) To start with, make the obvious observation that if a commuting square

$$\begin{array}{ccc} C & \subset & D \\ \cup & & \cup \\ A & \subset & B \end{array}$$

is described by the biunitary matrix u*, then the commuting square*

$$\begin{array}{ccc} B & \subset & D \\ \cup & & \cup \\ A & \subset & C \end{array}$$

is described by the biunitary matrix u^**.*

It follows, then, from Proposition 5.5.4(a) that if

$$\begin{array}{ccc} A_0^1 & \overset{H}{\subset} & A_1^1 \\ {\scriptstyle K}\cup & & \cup{\scriptstyle L} \\ A_0^0 & \overset{G}{\subset} & A_1^0 \end{array}$$

is a symmetric commuting square which is described by a biunitary matrix U*, and if we construct the grid* $\{A_k^n\}$ *starting from this initial commuting square in the usual manner, then the commuting squares*

$$\left(\begin{array}{ccc} A_1^1 & \overset{H'}{\subset} & A_2^1 \\ {\scriptstyle L}\cup & & \cup{\scriptstyle K} \\ A_1^0 & \overset{G'}{\subset} & A_2^0 \end{array}\right), \quad \left(\begin{array}{ccc} A_0^2 & \overset{G}{\subset} & A_1^2 \\ {\scriptstyle K'}\cup & & \cup{\scriptstyle L'} \\ A_0^1 & \overset{H}{\subset} & A_1^1 \end{array}\right) \quad \text{and} \quad \left(\begin{array}{ccc} A_1^2 & \overset{G'}{\subset} & A_2^2 \\ {\scriptstyle L'}\cup & & \cup{\scriptstyle K'} \\ A_1^1 & \overset{H'}{\subset} & A_2^1 \end{array}\right)$$

are described by the biunitary matrices $V, \tilde{V}$ and $\tilde{U}$ respectively, where V is as in Proposition 5.5.4, and

$$\begin{aligned} \tilde{v}^{\tilde{\kappa}\alpha}_{\beta\tilde{\lambda}} &= v^{\lambda\tilde{\beta}}_{\tilde{\alpha}\kappa}, \\ \tilde{u}^{\tilde{\lambda}\tilde{\alpha}}_{\tilde{\beta}\tilde{\kappa}} &= u^{\kappa\beta}_{\alpha\lambda}. \end{aligned}$$

Notice that $\tilde{\tilde{U}} = U$, and thus only these four biunitary matrices suffice to describe all the squares in the grid.

(ii) It follows from (i) above and equation (6.1.2) that if the matrix w gives rise to the subfactor $R_0 \subseteq R_1$ as in Theorem 6.1.4 or Theorem 6.1.5, then the transposed matrix w' gives rise to the dual subfactor $R_1 \subseteq R_2$.

6.2 Some examples

The input for a vertex model, as in Theorem 6.1.4 for instance, is an $Nk \times Nk$ matrix $w = ((w^{\alpha a}_{\beta b}))$, which satisfies the biunitarity condition that w is unitary, as also is the matrix v defined by $v^{\alpha a}_{b\beta} = \overline{w^{\beta a}_{\alpha b}}$; this is easily seen to be equivalent to the following condition:

Suppose we write w in block-form as $w = ((w^{\alpha}_{\beta}))_{1 \leq \alpha, \beta \leq N}$, where w^{α}_{β} is the $k \times k$ matrix defined by $(w^{\alpha}_{\beta})^{a}_{b} = w^{\alpha a}_{\beta b}$; then the biunitarity condition says that not only should w be unitary, but so also should be the matrix w'^{N} which is the matrix obtained by forming 'block-wise' transpose: i.e., in block-form, we have $(w'^{N})^{\alpha}_{\beta} = w^{\beta}_{\alpha}$.

EXAMPLE 6.2.1 *Let $\{\gamma_1, \gamma_2, \cdots, \gamma_N\}$ be any collection of $k \times k$ unitary matrices, and define*

$$w^{\alpha a}_{\beta b} = \left\{ \begin{array}{ll} (\gamma_\alpha)^{a}_{b} & \text{if } \alpha = \beta, \\ 0 & \text{otherwise.} \end{array} \right\} \tag{6.2.1}$$

It is trivially verified that this w satisfies the biunitarity condition stated above. (Reason: $w^{\alpha}_{\beta} = 0$ if $\alpha \neq \beta$, and hence what we have called w'^{N} is the same as w.) Notice that the 'diagonal constraint' above says that the Boltzmann weight associated to a crossing is zero unless the two ends of the over-strand have the same label.

It follows from Theorem 6.1.4 that $C_n = R'_0 \cap R_n$ consists of those matrices $F = ((F^{a_1 \cdots a_n}_{b_1 \cdots b_n})) \in \otimes^n M_k(\mathbb{C})$ for which there exists a matrix $G = ((G^{a_1 \cdots a_n}_{b_1 \cdots b_n})) \in \otimes^n M_k(\mathbb{C})$ such that

$$\begin{aligned} \delta^{\alpha}_{\beta} & \sum_{c_1, \cdots, c_n = 1}^{k} F^{a_1 \cdots a_n}_{c_1 \cdots c_n} (\gamma_\alpha)^{c_1}_{b_1} (\overline{\gamma_\alpha})^{c_2}_{b_2} (\gamma_\alpha)^{c_3}_{b_3} \cdots \\ = \delta^{\alpha}_{\beta} & \sum_{c_1, \cdots, c_n = 1}^{k} ((\gamma_\alpha)^{a_1}_{c_1} (\overline{\gamma_\alpha})^{a_2}_{c_2} (\gamma_\alpha)^{a_3}_{c_3} \cdots) G^{c_1 \cdots c_n}_{b_1 \cdots b_n} \end{aligned}$$

for all $a_1, \cdots, a_n, b_1, \cdots, b_n, \alpha, \beta$, *or equivalently,*

$$F(\underbrace{\gamma_\alpha \otimes \overline{\gamma_\alpha} \otimes \gamma_\alpha \otimes \cdots}_{n\ factors}) = (\underbrace{\gamma_\alpha \otimes \overline{\gamma_\alpha} \otimes \gamma_\alpha \otimes \cdots}_{n\ factors})G \ \ \forall \alpha.$$

If we let K *be the closed subgroup of* $U(k)$ *generated by* $\{\gamma_\alpha \gamma_\beta^{-1} : 1 \leq \alpha, \beta \leq N\}$, *and if we write* π *for the identity representation of* K *on* $\mathbb{C}^k$, *the previous conditions are seen to be entirely equivalent to*

$$C_n = (\underbrace{\pi \otimes \overline{\pi} \otimes \pi \otimes \cdots}_{n\ factors})(K)'.$$

In view of Remark 6.1.7, a moment's thought should convince the reader that the principal graph of this subfactor has the following description: first form a bipartite graph with the set of even (resp., odd) vertices being given by $\mathcal{G}^{(0)} = \hat{K} \times \{0\}$ *(resp.,* $\mathcal{G}^{(1)} = \hat{K} \times \{1\}$*) – where* $\hat{K}$ *denotes the unitary dual of the compact group* K *– where the number of bonds joining the vertices* $(\rho_i, i), i = 0, 1$, *is given by* $\langle \rho_0 \otimes \pi, \rho_1 \rangle$, *the mutiplicity with which* ρ_1 *features in the tensor product* $\rho_0 \otimes \pi$; *finally, the desired principal graph is the connected component of the above graph which contains the even vertex indexed by the trivial representation of* K.

Notice next that the transpose matrix w' *is given by the matrices* $\gamma_1', \cdots, \gamma_N'$ *in exactly the same way that* w *is given by the* γ_α*'s. Thus – in view of Remark 6.1.8(ii) – the dual graph is described by the closed subgroup* K' *of* $U(k)$ *– given by* $K' = \{g' : g \in K\}$ *– in the same way that the principal graph was described by* K. *Notice now that* $K' = \{\overline{g^{-1}} : g \in K\} = \{\overline{g} : g \in K\} = \overline{K}$. *Hence the equation* $\phi(g) = \overline{g}$ *defines an isomorphism* $\phi : K' \to K$. *Let* $\phi^* : \hat{K} \to \hat{K}'$ *denote the (obviously bijective) map defined by* $\phi^*(\rho) = \rho \circ \phi$. *Notice that if we write* π' *for the identity representation of* K' *on* $\mathbb{C}^k$, *then* $\pi' = \phi^*(\overline{\pi})$. *Finally, it follows that the principal graph is isomorphic to the dual graph by a graph isomorphism which associates the even or odd vertex in the former which is indexed by* ρ *(say) to the even or odd vertex in the latter which is indexed by* $\phi^*(\overline{\rho})$.

EXAMPLE 6.2.2 *Let* $\{c_1, c_2, \cdots, c_k\}$ *be any collection of* $N \times N$ *unitary matrices, and define*

$$w_{\beta b}^{\alpha a} = \left\{ \begin{array}{ll} (c_a)_\beta^\alpha & if\ a = b, \\ 0 & otherwise. \end{array} \right\} \tag{6.2.2}$$

It is trivially verified that this w *satisfies the biunitarity condition stated above. (Reason: each* w_β^α *is a diagonal matrix, and hence what we called* w'^N *is nothing but the transpose of* w.*) Notice that the 'diagonal constraint' above says that the Boltzmann weight associated to a crossing is zero unless the two ends of the under-strand have the same label.*

It follows from Theorem 6.1.4 that $C_n = R_0' \cap R_n$ *consists of those matrices* $F = ((F_{b_1 \cdots b_n}^{a_1 \cdots a_n})) \in \bigotimes^n M_k(\mathbb{C})$ *for which there exists a matrix* $G = ((G_{b_1 \cdots b_n}^{a_1 \cdots a_n})) \in$

$\bigotimes^n M_k(\mathbb{C})$ *such that*

$$\sum_{\gamma_1,\cdots,\gamma_{n-1}=1}^{N} F^{a_1\cdots a_n}_{b_1\cdots b_n} \underbrace{((c_{b_1})^{\alpha}_{\gamma_1}\overline{(c_{b_2})^{\gamma_2}_{\gamma_1}}(c_{b_3})^{\gamma_2}_{\gamma_3}\cdots)}_{n\ factors}$$
$$= \sum_{\gamma_1,\cdots,\gamma_{n-1}=1}^{N} \underbrace{((c_{a_1})^{\alpha}_{\gamma_1}\overline{(c_{a_2})^{\gamma_2}_{\gamma_1}}(c_{a_3})^{\gamma_2}_{\gamma_3}\cdots)}_{n\ factors} G^{a_1\cdots a_n}_{b_1\cdots b_n}$$

for all $a_1,\cdots,a_n,b_1,\cdots,b_n,\alpha,\beta$, *or equivalently*

$$F^{a_1\cdots a_n}_{b_1\cdots b_n}\underbrace{(c_{b_1}c^*_{b_2}c_{b_3}\cdots)}_{n\ factors}{}^{\alpha}_{\beta} = \underbrace{(c_{a_1}c^*_{a_2}c_{a_3}\cdots)}_{n\ factors}{}^{\alpha}_{\beta}G^{a_1\cdots a_n}_{b_1\cdots b_n}$$

for all $a_1,\cdots,a_n,b_1,\cdots,b_n,\alpha,\beta$; *since the 'variables separate', this shows that*

$$C_n = \{F = ((F^{a_1\cdots a_n}_{b_1\cdots b_n})) \in \bigotimes^n M_k(\mathbb{C}) : F^{a_1\cdots a_n}_{b_1\cdots b_n} = 0 \text{ unless}$$
$$c_{a_1}c^*_{a_2}c_{a_3}\cdots \text{ is a scalar multiple of } c_{b_1}c^*_{b_2}c_{b_3}\cdots\}.$$

After a moment's thought, this is seen to imply that the principal graph of this subfactor has the following description: let $\tilde{G}$ *be the (non-closed) subgroup of* $U_N(\mathbb{C})$ *generated by* $\{c_1,c_2,\cdots,c_k\}$ *and let* G *be the quotient of* $\tilde{G}$ *by the normal subgroup of those elements of* $\tilde{G}$ *which are scalar multiples of the identity matrix; form a bipartite graph with the sets of even and odd vertices being both indexed by* G*, where the number of bonds joining the even vertex indexed by* $[g_0]$ *to the odd vertex* $[g_1]$ *is given by the cardinality of the set* $\{1 \leq i \leq k : [g_1] = [g_0 c_i]\}$ *– where we have used the notation* $g \mapsto [g]$ *for the quotient mapping* $\tilde{G} \to G$*; finally, the desired principal graph is the connected component of the above graph which contains the even vertex indexed by the identity element of* $\tilde{G}$*.*

Here also, it is true that the dual graph is isomorphic to the principal graph. To see this, first note – in view of Remark 6.1.8(ii) – that w' *is constructed out of the* c'_a*'s in exactly the same manner that* w *was constructed out of the* c_a*'s. Using natural notations, we see that the dual graph is the Cayley graph of the group* $G' = \{g' : g \in G\}$ *modulo scalars, with respect to the generators* $\{[c'_a] : 1 \leq a \leq N\}$*. The group isomorphism* $g \mapsto g'$ *establishes the desired graph isomorphism.*

EXAMPLE 6.2.3 *We now discuss the second relative commutant of a subfactor constructed from a spin model (as in Theorem 6.1.5). Using the notation of that theorem, recall from §5.1 that* $R'_0 \cap R_2$ *consists of those* $F = ((F^a_b)) \in M_N(\mathbb{C})$ *such that there exists an array of numbers* $G = ((G_{b_1 b_2}))$ *(which should be actually thought of as an element of* C^{N^2}*) such that for all choices of* $a, b_1, b_2 \in \{1,2,\cdots,N\}$*,*

$$\sum_{x=1}^{N} F^a_x w^x_{b_1}\overline{w^x_{b_2}} = w^a_{b_1}\overline{w^a_{b_2}}G_{b_1 b_2}. \tag{6.2.3}$$

For each $\mathbf{a} = (a_1, a_2) \in \{1, 2, \cdots, N\} \times \{1, 2, \cdots, N\}$, *define the vector* $v_{\mathbf{a}} \in \mathbb{C}^N$ *by* $(v_{\mathbf{a}})_x = w^x_{a_1} \overline{w^x_{a_2}}$; *then equation (6.2.3) says that* $v_{\mathbf{b}}$ *is an eigenvector for the matrix* F *with eigenvalue* $G_{b_1 b_2}$. *Since* $R_0' \cap R_2$ *is generated by self-adjoint elements, and since eigenvectors of a self-adjoint matrix which correspond to distinct eigenvalues are orthogonal, we see that the matrix* G *must satisfy* $G_{b_1 b_2} = G_{a_1 a_2}$ *if* $v_{\mathbf{b}}$ *is not orthogonal to* $v_{\mathbf{a}}$. *More precisely, we can deduce the following: let* $\simeq_w$ *be the smallest equivalence relation in the set* $\{1, 2, \cdots, N\} \times \{1, 2, \cdots, N\}$ *such that* $\mathbf{a} \simeq_w \mathbf{b}$ *if* $v_{\mathbf{b}}$ *is not orthogonal to* $v_{\mathbf{a}}$. *Then any* G *as in (6.2.3) must necessarily be constant on equivalence classes. Conversely, if* G *is any matrix which is constant on equivalence classes, it is easily seen that the equation*

$$F^a_x = \sum_{b_1, b_2 = 1}^{N} G_{b_1 b_2} (v_{\mathbf{b}})_a \overline{(v_{\mathbf{b}})_x}$$

defines an element F *satisfying (6.2.3).*

It follows that $R_0' \cap R_2$ *is an abelian *-subalgebra of* $M_N(\mathbb{C})$ *with dimension equal to the number of* $\simeq_w$ *equivalence classes. It turns out – see [JNM] – that if* C *is one such equivalence class, then the number* $\#\{a : (a, i) \in C\}$ *is independent of* i*; call this number the* valency *of the equivalence class. It is a fact that if* F_C *is the minimal projection in* $R_0' \cap R_2$ *which is the projection onto the subspace spanned by* $\{v_{\mathbf{b}} : \mathbf{b} \in C\}$, *then the valency of* C *is precisely the rank of the projection* F_C.

On the other hand, since $\mathrm{tr}_{R_2}|_{R_0' \cap R_2} = \frac{1}{N} \mathrm{Tr}|_{R_0' \cap R_2}$, *where* Tr *denotes the (non-normalised) matrix-trace on* $M_N(\mathbb{C})$, *and since the subfactor determines not only the tower* $\{R_0' \cap R_n : n \geq 0\}$, *but also the traces* $\{\mathrm{tr}_{R_n}|_{R_0' \cap R_n} : n \geq 0\}$, *we see that the subfactor given by a spin model as above determines the number of* $\simeq_w$ *equivalence classes as well as the valencies of each of these components.*

We wish to make two points through this example:

(a) Spin models yield a passage from Hadamard matrices to subfactors via a commuting square. It is clear that two Hadamard matrices are equivalent – as described in §5.2.2 – precisely when the associated commuting squares are isomorphic (see Remark 5.5.3), and that in such a case, the associated subfactors are conjugate. The point here is that, via the preceding remarks, the subfactor associated to a Hadamard matrix w *in the above fashion determines the number of* $\simeq_w$ *equivalence classes and the valencies of those equivalence classes, and it turns out – see [JNM] – that these suffice to tell the five pairwise inequivalent real Hadamard matrices of order* 16 *from one another.*

(b) It is a fact that there exists a 16×16 *real Hadamard matrix which is not equivalent to its transpose. It follows from (a) above that the subfactor associated to this Hadamard matrix is not self-dual.*

6.3 On permutation vertex models

This section[1] is devoted to a brief discussion of vertex models for which the underlying biunitary matrix is a permutation matrix – i.e., has only 0's and 1's as entries. Recall – from the second paragraph of §6.2 – that a unitary matrix $w = ((w^{\alpha a}_{\beta b})) \in U(Nk)$ is biunitary precisely when its 'block-wise transpose' $\tilde{w}$, defined by $\tilde{w}^{\alpha a}_{\beta b} = w^{\beta a}_{\alpha b}$, is also unitary.

In the next few pages, we shall write $\tilde{w}$ for the block-wise transpose of w. Also we continue to use the conventions of this chapter concerning the use of Greek and Latin letters for elements of Ω_N and Ω_k respectively, where we write $\Omega_n = \{1, 2, \cdots, n\}$.

We shall find it convenient to work with an alternative description of such biunitary permutation matrices, which we single out in the next lemma.

LEMMA 6.3.1 *Let $w \in M_N(\mathbb{C}) \otimes M_k(\mathbb{C})$. Then the following conditions on w are equivalent:*

(i) w is biunitary, and is further a permutation matrix (i.e., is a $0,1$ matrix);

(ii) there exist permutations $\{\rho_a : a \in \Omega_k\} \subset S(\Omega_N), \{\lambda_\alpha : \alpha \in \Omega_N\} \subset S(\Omega_k)$ (where we write $S(X)$ for the group of permutations of the set X), such that:

(a) the equation

$$\pi(\beta, b) = (\rho_b(\beta), \lambda_\beta(b))$$

defines a permutation $\pi \in S(\Omega_N \times \Omega_k)$; and

(b)

$$w^{\alpha a}_{\beta b} = \delta_{(\alpha,a),\pi(\beta,b)} = \delta_{\alpha,\rho_b(\beta)}\delta_{a,\lambda_\beta(b)}.$$

Proof: (i) $\Rightarrow$ (ii): If w is a biunitary $0,1$-valued matrix, then let $\pi \in S(\Omega_N \times \Omega_k)$ be defined by $w^{\alpha a}_{\beta b} = \delta_{(\alpha,a),\pi(\beta,b)}$.

Assertion: For any $\beta \in \Omega_N, a \in \Omega_k$ (resp., $\alpha \in \Omega_N, b \in \Omega_k$), $\pi(\{\beta\} \times \Omega_k) \cap (\Omega_N \times \{a\})$ (resp., $\pi(\Omega_N \times \{b\}) \cap (\{\alpha\} \times \Omega_k)$) is a singleton. Furthermore,

$$\begin{aligned} \pi(\{\beta\} \times \Omega_k) \cap (\Omega_N \times \{a\}) &= \{(\phi_a(\beta), a)\}, \\ \pi(\Omega_N \times \{b\}) \cap (\{\alpha\} \times \Omega_k) &= \{(\alpha, \psi_\alpha(b))\}, \end{aligned}$$

where

$$\phi_a(\beta) = \rho_{\lambda_\beta^{-1}(a)}(\beta) \text{ and } \psi_\alpha(b) = \lambda_{\rho_b^{-1}(\alpha)}(b).$$

The first (as well as the parenthetically included) statement of the assertion is an immediate consequence of two facts: (i) the hypothesis on w

[1]This section is a reproduction, almost verbatim in places, of parts of the paper [KS].

implies that the block-transpose matrix $\tilde{w}$ is also a permutation matrix; and (ii) $\pi(\beta, b) = (\alpha, a) \Leftrightarrow \tilde{w}^{\beta a}_{\alpha b} = 1$.

The second assertion follows from the definitions.

The assertion clearly proves the implication (i) $\Rightarrow$ (ii), while the implication (ii) $\Rightarrow$ (i) is immediate. □

We shall find the following notation convenient.

DEFINITION 6.3.2 *Define*

$$P_{N,k} = \{\pi \in S(\Omega_N \times \Omega_k) : \text{there exists } \lambda : \Omega_N \to S(\Omega_k), \rho : \Omega_k \to S(\Omega_N) \\ \text{such that } \pi(\beta, b) = (\rho_b(\beta), \lambda_\beta(b)) \text{ for all } \beta \in \Omega_N, b \in \Omega_k\}$$

where we write λ_β (resp.,ρ_b) for the image of β (resp.,b) under the map λ (resp.,ρ). If π, λ, ρ are related as above, we shall simply write $\pi \leftrightarrow (\rho, \lambda) \in P_{N,k}$.

Thus Lemma 6.3.1 states that there is a bijection between biunitary permutation matrices of size Nk and elements $\pi \leftrightarrow (\rho, \lambda) \in P_{N,k}$, given by $w^{\alpha a}_{\beta b} = \delta_{\alpha,\rho_b(\beta)}\delta_{a,\lambda_\beta(b)}$.

The following proposition, which is an immediate consequence of the definitions, lists some useful properties of the various ingredients of a biunitary permutation.

PROPOSITION 6.3.3 *Let $\pi \leftrightarrow (\rho, \lambda), \phi_a, \psi_\alpha$ be as above. Then, for arbitrary $a \in \Omega_k, \alpha \in \Omega_N$, we have:*

(i) $\phi_a \in S(\Omega_N), \psi_\alpha \in S(\Omega_k)$;

(ii) $\pi^{-1} \leftrightarrow (\phi^{-1}, \psi^{-1}) \in P_{N,k}$ (meaning, of course, that $\pi^{-1}(\alpha, a) = (\phi_a^{-1}(\alpha), \psi_\alpha^{-1}(a))$);

(iii) $\phi_a^{-1}(\alpha) = \rho^{-1}_{\psi_\alpha^{-1}(a)}(\alpha), \psi_\alpha^{-1}(a) = \lambda^{-1}_{\phi_a^{-1}(\alpha)}(a)$. □

For the rest of this section, we fix a $\pi \leftrightarrow (\rho, \lambda) \in P_{N,k}$ and let $\lambda, \rho, \phi, \psi$ be as above. Thus, if w is the biunitary permutation matrix that corresponds to (ρ, λ), then

$$w^{\alpha a}_{\beta b} = \delta_{\alpha,\rho_b(\beta)}\delta_{a,\lambda_\beta(b)}. \tag{6.3.1}$$

The point of the next lemma is to point out that if $w^{\alpha a}_{\beta b} = 1$, then any pair consisting of one Greek letter from $\{\alpha, \beta\}$ and one Latin letter from $\{a, b\}$ determines the complementary pair. We shall find some of these formulae convenient in subsequent computations.

LEMMA 6.3.4 *If $\alpha, \beta \in \Omega_N, a, b \in \Omega_k$, then the following conditions are equivalent:*

(i) $w^{\alpha b}_{\beta a} = 1$;

(ii) $\alpha = \rho_a(\beta)$ and $b = \lambda_\beta(a)$;

(iii) $\beta = \phi_b^{-1}(\alpha)$ *and* $a = \psi_\alpha^{-1}(b)$;

(iv) $\beta = \rho_a^{-1}(\alpha)$ *and* $b = \psi_\alpha(a)$;

(v) $\alpha = \phi_b(\beta)$ *and* $a = \lambda_\beta^{-1}(b)$.

Proof: (i) ⇔ (ii) by definition.
(ii) ⇔ (iii) by Proposition 6.3.3(ii).
(ii) ⇔ (iv) by the formula for ϕ^{-1} given in Proposition 6.3.3(iii).
(iii) ⇔ (v) by the formula for ψ^{-1} given in Proposition 6.3.3(iii). □

We wish to discuss the higher relative commutants $C_n = R_w' \cap R_{n-1}, n \geq 0$, where $R_w = R_{-1} \subseteq R = R_0 \subseteq R_1 \subseteq \cdots \subseteq R_n \subseteq \cdots$ is the tower associated to the subfactor $R_w \subseteq R$ constructed from the commuting square given by w in the usual manner. Before we get to that, notice the following consequence of the preceding lemma: the Boltzmann weights associated with the two kinds of crossings (as per the prescription of §6.1) are as follows:

Positive Crossing

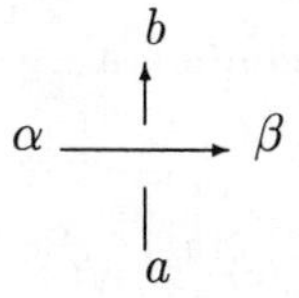

$$\mapsto \quad \delta_{\beta,\rho_a^{-1}(\alpha)}\delta_{b,\psi_\alpha(a)} = \delta_{\alpha,\phi_b(\beta)}\delta_{a,\lambda_\beta^{-1}(b)}$$

Negative Crossing

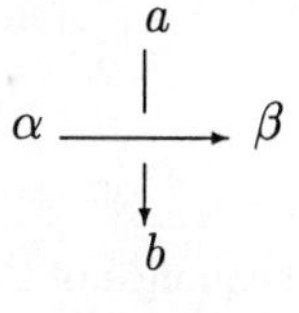

$$\mapsto \quad \delta_{\beta,\rho_b(\alpha)}\delta_{a,\lambda_\alpha(b)} = \delta_{\alpha,\phi_a^{-1}(\beta)}\delta_{b,\psi_\beta^{-1}(a)} \tag{6.3.2}$$

Notation: Given a biunitary permutation w and corresponding maps λ, ρ, ϕ, ψ as above, then for arbitrary $n \geq 1$ and $\mathbf{a} \in \Omega_k^n$, we define the alternating products

$$\rho_{\mathbf{a}} = \rho_{a_1}\rho_{a_2}^{-1}\rho_{a_3}\cdots\rho_{a_n}^{\pm}$$

and

$$\phi_{\mathbf{a}} = \phi_{a_1}\phi_{a_2}^{-1}\phi_{a_3}\cdots\phi_{a_n}^{\pm}.$$

We are now ready to introduce certain mappings that will play a central role in the computation of the higher relative commutants.

PROPOSITION 6.3.5 *(i) For all* $n \geq 1$, *there exists a mapping* $\Omega_N \ni \alpha \mapsto$

$L_\alpha^{(n)} \in S(\Omega_k^n)$ *such that*

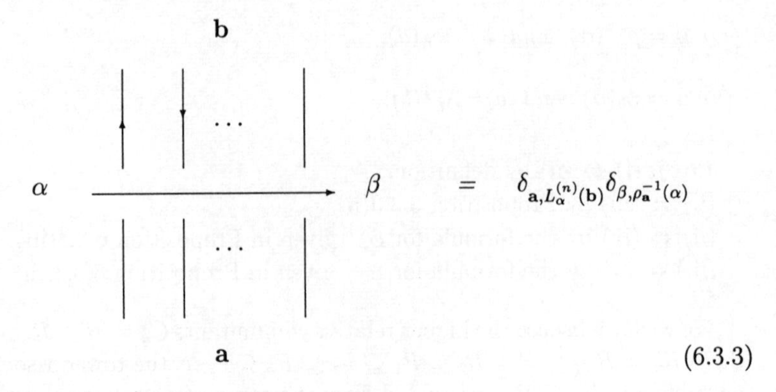

$$= \delta_{\mathbf{a}, L_\alpha^{(n)}(\mathbf{b})} \delta_{\beta, \rho_{\mathbf{a}}^{-1}(\alpha)} \qquad (6.3.3)$$

where the $L_\alpha^{(n)}$*'s are defined as in (ii) below.*

(ii) $L_\alpha^{(1)} = \psi_\alpha^{-1}$; *and if* $n > 1$ *and if* $L_\alpha^{(n)}(\mathbf{b}) = \mathbf{a}$, *then*

$$\mathbf{a}_{n-1]} = L_\alpha^{(n-1)}(\mathbf{b}_{n-1]})$$

(where we have used the obvious notation $\mathbf{a}_{n-1]}$ *to mean* $(a_1, \cdots, a_{n-1})$ *if* $\mathbf{a} = (a_1, \cdots, a_n)$*); and*

$$a_n = \begin{cases} \lambda^{-1}_{\phi^{-1}_{\mathbf{b}_{n-1]}}(\alpha)}(b_n) & \text{if } n \text{ is even,} \\ \psi^{-1}_{\phi^{-1}_{\mathbf{b}_{n-1]}}(\alpha)}(b_n) & \text{if } n \text{ is odd.} \end{cases}$$

(iii) $\rho^{-1}_{L_\alpha^{(n)}(\mathbf{b})}(\alpha) = \phi_{\mathbf{b}}^{-1}(\alpha)$.

Proof: The proof is a direct consequence of the prescription, given in equation (6.3.2), for the Boltzmann weights associated to positive and negative crossings. (For (iii), the two prescriptions given for each kind of crossing must be used in conjunction.) □

In the following, we fix a biunitary w, with associated $\lambda, \rho, \phi, \psi$ as above, and let $\{C_n\}$ denote the sequence of higher relative commutants for this R_w.

Lemma 6.3.6 *With the identification* $\otimes^n M_k(\mathbb{C}) = \mathrm{Mat}_{\Omega_k^n}(\mathbb{C})$, *we have*

$$C_n = \{F = ((F_{\mathbf{b}}^{\mathbf{a}})) \in \mathrm{Mat}_{\Omega_k^n}(\mathbb{C}) : F_{\mathbf{b}}^{\mathbf{a}} = \delta_{\rho_{L_\alpha^{(n)}(\mathbf{a})}, \rho_{L_\alpha^{(n)}(\mathbf{b})}} F^{L_\beta^{(n)^{-1}} L_\alpha^{(n)}(\mathbf{a})}_{L_\beta^{(n)^{-1}} L_\alpha^{(n)}(\mathbf{b})} \text{ for all } \alpha, \beta \in \Omega_N, \mathbf{a}, \mathbf{b} \in \Omega_k^n\}.$$

Proof: In the notation of Theorem 6.1.4, we see, from Proposition 6.3.5,

that on the one hand,

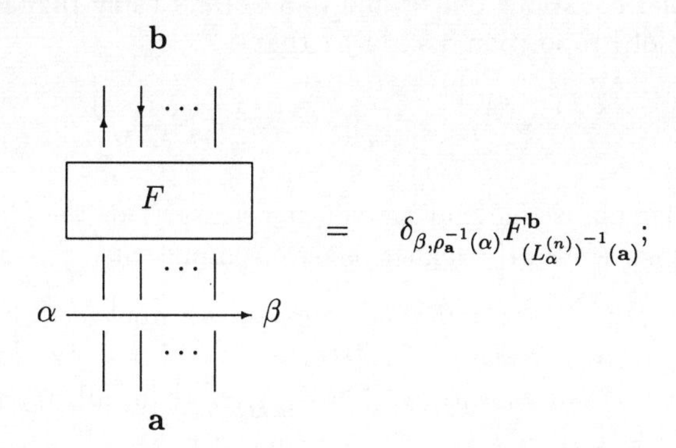

$$= \delta_{\beta,\rho_{\mathbf{a}}^{-1}(\alpha)} F^{\mathbf{b}}_{(L_\alpha^{(n)})^{-1}(\mathbf{a})};$$

on the other hand, we also find that

$$= \delta_{\beta,\rho^{-1}_{L_\alpha^{(n)}(\mathbf{b})}(\alpha)} G_{\mathbf{a}}^{L_\alpha^{(n)}(\mathbf{b})},$$

which, in view of Proposition 6.3.5(iii), is seen to be equal to $\delta_{\beta,\phi_{\mathbf{b}}^{-1}(\alpha)} G_{\mathbf{a}}^{L_\alpha^{(n)}(\mathbf{b})}$.

Thus we find that C_n consists of those $F \in \bigotimes^n M_k(\mathbb{C})$ for which there exists a $G \in \bigotimes^n M_k(\mathbb{C})$ such that

$$\delta_{\beta,\rho_{\mathbf{a}}^{-1}(\alpha)} F^{\mathbf{b}}_{(L_\alpha^{(n)})^{-1}(\mathbf{a})} = \delta_{\beta,\phi_{\mathbf{b}}^{-1}(\alpha)} G_{\mathbf{a}}^{L_\alpha^{(n)}(\mathbf{b})}.$$

Using the substitution $\mathbf{c} = (L_\alpha^{(n)})^{-1}(\mathbf{a})$, the last equation may be re-written – again using Proposition 6.3.5(iii) – in a more symmetric form as

$$\delta_{\beta,\phi_{\mathbf{c}}^{-1}(\alpha)} F^{\mathbf{b}}_{\mathbf{c}} = \delta_{\beta,\phi_{\mathbf{b}}^{-1}(\alpha)} G^{L_\alpha^{(n)}(\mathbf{b})}_{L_\alpha^{(n)}(\mathbf{c})}. \tag{6.3.4}$$

This is easily seen to imply that

$$F^{\mathbf{b}}_{\mathbf{c}} = \delta_{\phi_{\mathbf{b}}^{-1},\phi_{\mathbf{c}}^{-1}} G^{L_\alpha^{(n)}(\mathbf{b})}_{L_\alpha^{(n)}(\mathbf{c})} \tag{6.3.5}$$

for arbitrary $\alpha \in \Omega_N$, $\mathbf{b}, \mathbf{c} \in \Omega_k^n$, and also (as a result of Proposition 6.3.5(iii)) that

$$G^{\tilde{\mathbf{b}}}_{\tilde{\mathbf{c}}} = \delta_{\rho_{\tilde{\mathbf{b}}}^{-1},\rho_{\tilde{\mathbf{c}}}^{-1}} F^{(L_\beta^{(n)})^{-1}(\tilde{\mathbf{b}})}_{(L_\beta^{(n)})^{-1}(\tilde{\mathbf{c}})} \tag{6.3.6}$$

for arbitrary $\beta \in \Omega_N, \tilde{\mathbf{b}}, \tilde{\mathbf{c}} \in \Omega_k^n$. The proof of the lemma is completed by putting together equations (6.3.5) and (6.3.6) (and using the fact – which is a consequence of Proposition 6.3.5(iii) – that

$$\delta_{\phi_{\mathbf{b}}^{-1},\phi_{\mathbf{c}}^{-1}}\delta_{\rho^{-1}_{L_\alpha^{(n)}(\mathbf{b})},\rho^{-1}_{L_\alpha^{(n)}(\mathbf{c})}} = \delta_{\rho^{-1}_{L_\alpha^{(n)}(\mathbf{b})},\rho^{-1}_{L_\alpha^{(n)}(\mathbf{c})}}). \qquad \Box$$

The next lemma is the final ingredient necessary for the identification – in an abstract sense – of the higher relative commutants.

LEMMA 6.3.7 *Let Ω be a finite set. Suppose we are given an equivalence relation $\sim_0$ on Ω and a subset $\mathcal{L} \subseteq S(\Omega)$ such that $\mathcal{L} = \mathcal{L}^{-1} = \{\sigma^{-1} : \sigma \in \mathcal{L}\}$. Let $A = \{x = ((x_j^i)) \in \mathrm{Mat}_\Omega(\mathbb{C}) : x_j^i = \delta_{[i]_0,[j]_0} x_{\sigma(j)}^{\sigma(i)}$ for all $i, j \in \Omega, \sigma \in \mathcal{L}\}$, where $[i]_0 = \{j \in \Omega : i \sim_0 j\}$.*

Define the equivalence relation $\sim$ on Ω by requiring that $i \sim j \Leftrightarrow \sigma(i) \sim_0 \sigma(j)$ for all $\sigma \in G$, where G is the subgroup of $S(\Omega)$ generated by $\mathcal{L}$. Then G acts on the set of $\sim$-equivalence classes (by $\sigma \cdot [i] = [\sigma(i)]$, where of course $[i] = \{j \in \Omega : i \sim j\}$). Suppose the set of $\sim$-equivalence classes breaks up as a disjoint union of l orbits under this action of G.

For $1 \le p \le l$, fix one equivalence class $[i_p]$ from the p-th orbit, let $H_p = \{\sigma \in G : \sigma \cdot [i_p] = [i_p]\}$ be the isotropy group of that equivalence class, and let π_p denote the natural permutation representation of H_p on $[i_p]$. Then

$$A \cong \bigoplus_{p=1}^{l} \pi_p(H_p)'.$$

Proof: To begin with, if $\sigma_1, \sigma_2 \in \mathcal{L}$, note that, for any x in A, we have $x_j^i = \delta_{[i]_0,[j]_0} x_{\sigma_2(j)}^{\sigma_2(i)} = \delta_{[i]_0,[j]_0}\delta_{[\sigma_2(i)]_0,[\sigma_2(j)]_0} x_{\sigma_1\sigma_2(j)}^{\sigma_1\sigma_2(i)}$. Since $\mathcal{L} = \mathcal{L}^{-1}$, clearly $G = \{\sigma_1\sigma_2\cdots,\sigma_r : r \ge 0, \sigma_1,\cdots,\sigma_r \in \mathcal{L}\}$, and it easily follows now that

$$A = \{x = ((x_j^i)) \in \mathrm{Mat}_\Omega(\mathbb{C}) : x_j^i = \delta_{[i],[j]} x_{\sigma(j)}^{\sigma(i)} \ \forall i, j \in \Omega, \sigma \in G\}.$$

Suppose now that $\{[j_1^{(p)}], \cdots [j_{t_p}^{(p)}]\}$ is the p-th orbit in the set of $\sim$-equivalence classes under the G-action, and suppose $j_1^{(p)} = i_p$. For $1 \le s \le t_p$, fix $\sigma_s^{(p)} \in G$ such that $\sigma_s^{(p)} \cdot [i_p] = [j_s^{(p)}]$. Assume that the elements of Ω have been so ordered that $\sigma_s^{(p)}$, as a map of $[j_1^{(p)}]$ onto $[j_s^{(p)}]$, is order-preserving. It is then fairly easy to see that $x \in A$ if and only if x has the block-diagonal form $x = \bigoplus_{p=1}^{l}\bigoplus_{s=1}^{t_p} x_s^{(p)}$ with respect to the decomposition $\Omega = \coprod_{p=1}^{l}\coprod_{s=1}^{t_p}[j_s^{(p)}]$, where $x_1^{(p)} = \cdots = x_{t_p}^{(p)} \in \pi_p(H_p)'$. $\Box$

Putting the previous two lemmas together – by considering the specialisation of Lemma 6.3.7 to the case where $\Omega = \Omega_k^n, \mathbf{a} \sim_0 \mathbf{b} \Leftrightarrow \rho_{L_\alpha^{(n)}(\mathbf{a})} = \rho_{L_\alpha^{(n)}(\mathbf{b})} \ \forall \alpha \in \Omega_N$, and $\mathcal{L} = \{(L_\beta^{(n)})^{-1}(L_\alpha^{(n)}) : \alpha, \beta \in \Omega_N\}$ – we can summarise the contents of this section as follows:

PROPOSITION 6.3.8 *Let $w \in M_N(\mathbb{C}) \otimes M_k(\mathbb{C})$ be a biunitary permutation matrix and let $\lambda, \rho, \phi, \psi$ have their usual meaning. Let $R_w \subseteq R$ be the hyperfinite (subfactor, factor)-pair corresponding to w, and let $C_n = R_w' \cap R_{n-1}, n \geq 0$, where $R_w = R_{-1} \subseteq R = R_0 \subseteq R_1 \subseteq R_2 \subseteq \cdots$ is the tower of the basic construction. Then, for $n = 1, 2, \cdots$, the algebra C_n has the following description:*

Let $L_\alpha^{(n)}$ be defined as in Proposition 6.3.5; let G_n be the subgroup of $S(\Omega_k^n)$ generated by $\{L_\beta^{(n)^{-1}} L_\alpha^{(n)} : \alpha, \beta \in \Omega_N\}$; and let $\sim_n$ be the equivalence relation defined on Ω_k^n by

$$\mathbf{a} \sim_n \mathbf{b} \Leftrightarrow \rho_{L_\alpha^{(n)}(\sigma(\mathbf{a}))} = \rho_{L_\alpha^{(n)}(\sigma(\mathbf{b}))} \quad \forall \sigma \in G_n, \alpha \in \Omega_N.$$

Suppose the set of equivalence classes in Ω_k^n breaks up into l_n orbits under the G_n-action; fix an equivalence class $[\alpha_p]$ in the p-th orbit of equivalence classes, and let $H_p = \{\sigma \in G_n : \sigma(\alpha_p) \sim_n \alpha_p\}$. If π_p is the natural permutation representation of H_p on $[\alpha_p]$, then

$$C_n \simeq \bigoplus_{p=1}^{l_n} \pi_p(H_p)'. \qquad \square$$

We now discuss a few simple special examples.

EXAMPLE 6.3.9 *(a) First consider the trivial case $\lambda \equiv \mathrm{id}_{\Omega_k}$, $\rho \equiv \mathrm{id}_{\Omega_N}$. In this most trivial example, $\pi(\beta, b) = (\beta, b)$ and in this case, the subfactor R_w of $R = M_k(\mathbb{C}) \otimes (\bigotimes_{n=1}^{\infty} M_N(\mathbb{C}))$ may be identified with $1 \otimes \bigotimes_{n=1}^{\infty} M_N(\mathbb{C})$, and the principal graph consists of two vertices with k bonds between them.*

(b) Let $\lambda \equiv \mathrm{id}_{\Omega_k}$, and let $\rho : \Omega_k \to S(\Omega_N)$ be an arbitrary map. Then $\pi(\beta, b) = (\rho_b(\beta), b)$, which clearly defines a permutation of $\Omega_N \times \Omega_k$; i.e., $\pi \leftrightarrow (\rho, \lambda) \in P_{N,k}$. Then observe that

$$\phi_a(\alpha) = \rho_{\lambda_\alpha^{-1}(a)}(\alpha) = \rho_a(\alpha), \psi_\alpha(a) = \lambda_{\rho_a^{-1}(\alpha)}(a) = a = \lambda_\alpha(a)$$

and thus $\phi = \rho, \psi = \lambda$. It follows from Proposition 6.3.5(ii) that, for all $n \geq 1$,

$$L_\alpha^{(n+1)}(\mathbf{a}, a) = (L_\alpha^{(n)}(\mathbf{a}), \lambda_{\rho_{\mathbf{a}}^{-1}(\alpha)}^{-1}(a)) = (L_\alpha^{(n)}(\mathbf{a}), a),$$

for all $\alpha \in \Omega_N, a \in \Omega_k, \mathbf{a} \in \Omega_k^n, n \geq 1$; hence, inductively, we find that $L_\alpha^{(n)} = \mathrm{id}_{\Omega_k^n}$ for all $n \geq 1$ and for all $\alpha \in \Omega_N$. In this case, the equivalence classes of Ω_k^n are the sets $E_\sigma = \{\mathbf{a} \in \Omega_k^n : \rho_{\mathbf{a}} = \sigma\}$, as σ ranges over the group G_0 generated by $\{\rho_i : i \in \Omega_N\}$. (Actually, E_σ is empty unless σ has the form $\rho_{a_1}\rho_{a_2}^{-1}\rho_{a_3}\cdots$.)

In fact, the hypothesis implies that $w_{\beta b}^{\alpha a} = \delta_{a,b}\delta_{\alpha,\rho_b(\beta)}$, and hence this case comes under the purview of the vertex models discussed in Example 6.2.2.

(c) Let $\lambda : \Omega_N \to S(\Omega_k)$ be an arbitrary map and let $\rho_a = \mathrm{id}_{\Omega_N}$ for all a; thus, $\pi(\beta, b) = (\beta, \lambda_b(\beta))$, which is again clearly a permutation of $\Omega_N \times \Omega_k$,

whence $\pi \leftrightarrow (\rho, \lambda) \in P_{N,k}$. *Observe again that* $\phi_a(\alpha) = \rho_{\lambda_\alpha^{-1}(a)}(\alpha) = \alpha = \rho_a(\alpha)$ *and that* $\psi_\alpha(a) = \lambda_{\rho_a^{-1}(\alpha)}(a) = \lambda_\alpha(a)$, *so that* $\phi = \rho, \psi = \lambda$. *It follows, again from Proposition 6.3.5(ii), that*

$$L_\alpha^{(n)} = \lambda_\alpha^{-1} \times \lambda_\alpha^{-1} \times \cdots \times \lambda_\alpha^{-1}.$$

We assume, for simplicity, that $\lambda_1 = \mathrm{id}$. *Then, if* G_1 *denotes the subgroup of* $S(\Omega_k)$ *generated by* $\{\lambda_\alpha : \alpha \in \Omega_N\}$, *we see, in the notation of Proposition 6.3.8, that* $G_k = \{\sigma \times \sigma \times \cdots \times \sigma : \sigma \in G_1\}$, *that the equivalence relation on* Ω_N^k *is the trivial one* ($\alpha \sim \beta$ *for all* α, β) *– as a result of the triviality of the* ρ_i*'s – and if* π *denotes the natural representation of* G_1 *on* $\mathbb{C}^N$, *then* $C_k \cong (\pi \otimes \pi \otimes \cdots \otimes \pi)(G_1)'$.

Again, in this case, we have $w_{\beta b}^{\alpha a} = \delta_{\alpha,\beta}\delta_{a,\lambda_\beta(b)}$, *so this case falls under the purview of Example 6.2.1.*

(d) We may obtain the tensor product of cases (b) and (c) above, by the following device: if $\lambda^{(1)} : \Omega_{N_1} \to S(\Omega_{k_1})$ *and* $\rho^{(2)} : \Omega_{k_2} \to S(\Omega_{N_2})$ *are arbitrary maps, set* $N = N_1N_2, k = k_1k_2$, *and define* $\lambda_\alpha, \rho_{\mathbf{a}}$ *by*

$$\lambda_\alpha(\mathbf{a}) = (\lambda_{\alpha_1}^{(1)}(a_1), a_2), \rho_{\mathbf{a}}(\alpha) = (\alpha_1, \rho_{a_2}^{(2)}(\alpha_2)).$$

(We have made the obvious identification $\Omega_N = \Omega_{N_1} \times \Omega_{N_2}, \Omega_k = \Omega_{k_1} \times \Omega_{k_2}$, *and denoted a typical element of* Ω_N *(resp.,* Ω_k*) by* $\alpha = (\alpha_1, \alpha_2)$ *(resp.,* $\mathbf{a} = (a_1, a_2)$*).)*

Lest the reader should get the wrong impression that examples obtained from permutation vertex models are all 'trivial' in some sense, we should mention that already when $N = k = 3$, there exists a permutation vertex model whose associated subfactor is irreducible, has infinite depth and has no 'intermediate subfactors'. The interested reader may consult [KS] for the details.

In fact, the reason for including this section here is that these permutation models are a potential source of interesting subfactors.

6.4 A diagrammatic formulation

In this section, we discuss a diagrammatic formulation – along the lines of our discussion of vertex models and spin models earlier in this chapter – that is valid for a general non-degenerate commuting square (with respect to the Markov trace).

We assume throughout this section that

$$\begin{array}{ccc} B_0 & \overset{L}{\subset} & B_1 \\ {\scriptstyle G}\cup & & \cup{\scriptstyle H} \\ A_0 & \overset{K}{\subset} & A_1 \end{array} \tag{6.4.1}$$

is a non-degenerate commuting square with respect to the Markov trace, with inclusions as indicated, where we assume that all the inclusions are connected. Further, we shall use the notation $\mathcal{G}, \mathcal{H}, \mathcal{K}$ and $\mathcal{L}$ to denote the Bratteli diagrams encoded by the inclusion matrices G, H, K and L, respectively; also, as before, we shall denote typical edges in the graphs $\mathcal{G}, \mathcal{H}, \mathcal{K}$ and $\mathcal{L}$ by α, β, κ and λ respectively.

Let us write $A_0^0 = A_0, A_1^0 = B_0, A_0^1 = A_1,$ and $A_1^1 = B_1$, so that the commuting square (6.4.1), when 'transposed', looks like this:

$$\begin{array}{ccc} A_0^1 & \overset{H}{\subset} & A_1^1 \\ K\cup & & \cup L \\ A_0^0 & \overset{G}{\subset} & A_1^0. \end{array} \tag{6.4.2}$$

Let $\{A_n^1 : n \geq 0\}$ be the tower of the basic construction for the initial inclusion $A_0^1 \subseteq A_1^1$, with the projection implementing the conditional expectation of A_n^1 onto A_{n-1}^1 being given, as usual, by e_{n+1}, for $n \geq 1$. As usual, let $A_n^0 = \langle A_{n-1}^0, e_n \rangle, n > 1$.

If $R_0 \subseteq R_1$ is the subfactor constructed out of the commuting square (6.4.1) by iterating the basic construction in the usual fashion, then it follows from the analysis of §5.7 that the higher relative commutants are given by

$$R_0' \cap R_n = A_0^{1'} \cap A_n^0, n > 0.$$

Once and for all, let us fix a biunitary matrix $u = ((u_{\alpha\circ\lambda}^{\kappa\circ\beta}))$ which describes the commuting square (6.4.2). Let us simply write $\bar{t}$ for the trace-vector on each $A_k^n, 0 \leq n, k \leq 1$. (Thus $\bar{t} : \bigcup_{n,k=0}^1 \pi(A_k^n) \to [0,1], H'H(\bar{t}|\pi(A_1^1)) = ||H||^2\bar{t}|\pi(A_1^1)$, etc.)

We begin by describing how to represent elements of $A_n^0, n \geq 0$. Actually, since we are only interested in relative commutants, we shall only discuss elements of $A_0^{0'} \cap A_n^0$. We shall think of a typical element of this relative commutant as a 'black box' with two sets of n strands, thus:

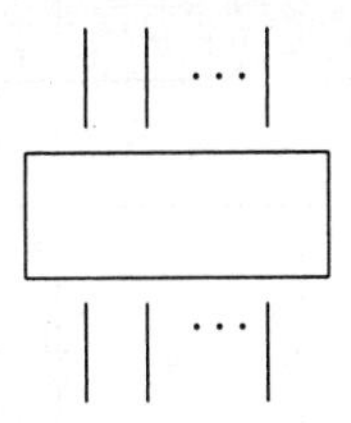

Such a black box is thought of as a scalar-valued function on the set of possible *states*, where a state (for such a simple diagram) is an assignment of vertices (from $\bigcup_{k=0}^1 \pi(A_k^0)$) to the regions, and edges (from $\mathcal{G}$) to the strands of the box, in such a way that (a) the assignment is a 'graph-theoretic homomorphism' (meaning that if a strand is labelled by an edge α, and if the two regions adjacent to the strand are labelled by vertices v_1, v_2 of $\mathcal{G}$, then α

must be an edge from v_1 to v_2 in $\mathcal{G}$), and (b) the region at the extreme left is labelled by a vertex from $\pi(A_0^0)$. (We think of the Bratteli diagram as an unoriented graph in this section.)

Thus a state – when $n = 2$ – may be given thus:

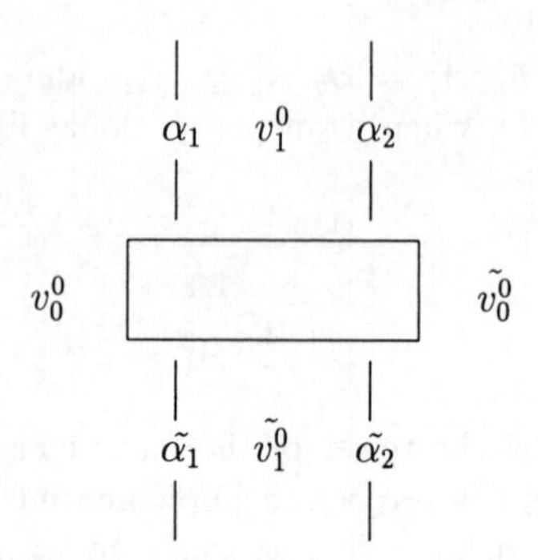

where $v_0^0, \tilde{v_0^0} \in \pi(A_0^0), v_1^0, \tilde{v_1^0} \in \pi(A_1^0), \alpha_1$ (resp., $\tilde{\alpha_1}$) is an edge joining v_0^0 to v_1^0 (resp., $\tilde{v_1^0}$) and α_2 (resp., $\tilde{\alpha_2}$) is an edge joining v_1^0 (resp., $\tilde{v_1^0}$) to $\tilde{v_0^0}$.

The description of elements of $A_0^{0\prime} \cap A_n^1$ is similar except for the following variations: here, there are two sets of $n+1$ strands, and a state should label the first strands (from the left) by edges in the Bratteli diagram $\mathcal{K}$, and all subsequent strands should be labelled by edges from the Bratteli diagram $\mathcal{H}$; as to the regions, the region on the extreme left should be labelled by a vertex from $\pi(A_0^0)$, and subsequent regions should be labelled by vertices from $\bigcup_{k=0}^1 \pi(A_k^1)$, and the 'homomorphic property' should be preserved; thus, a state, in this case, might look like this (when $n = 1$):

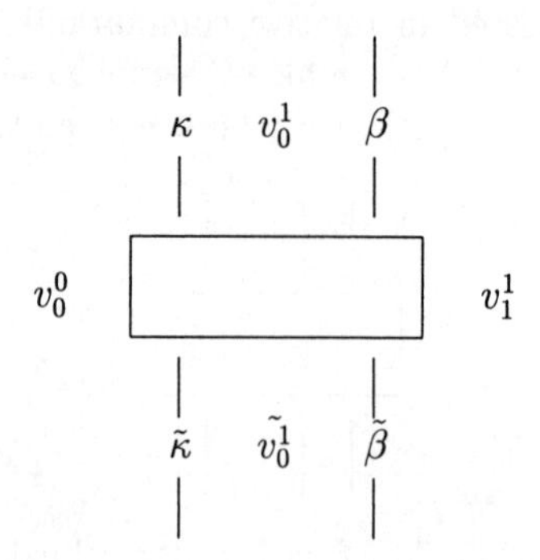

where, of course, $v_0^0 \in \pi(A_0^0), v_0^1, \tilde{v_0^1} \in \pi(A_0^1), v_1^1 \in \pi(A_1^1), \kappa$ (resp., $\tilde{\kappa}$) is an edge joining v_0^0 to v_0^1 (resp., $\tilde{v_0^1}$) in $\mathcal{K}$, and β (resp., $\tilde{\beta}$) is an edge joining v_0^1 (resp., $\tilde{v_0^1}$) to v_1^1 in $\mathcal{H}$.

The inclusion of $(A_0^{0\prime} \cap A_n^k)$ in $(A_0^{0\prime} \cap A_{n+1}^k)$ is given by the obvious identification:

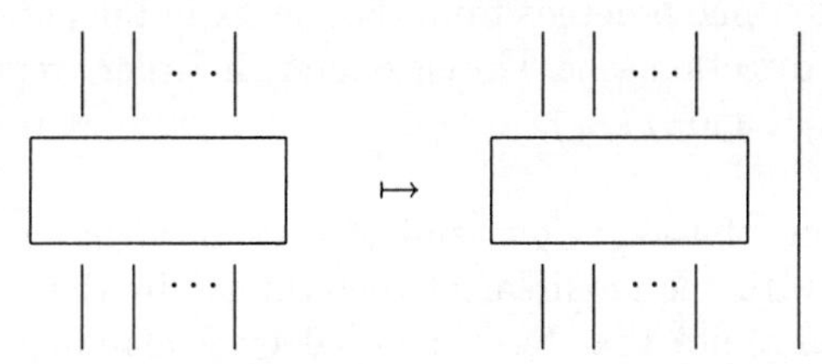

The (sometimes more complex) diagrams we shall be working with will have two other components in addition to (zero or one or many) black boxes, these being: (a) local extrema; and (b) crossings. Before we get to discussing these, we pause to mention a few features of the diagrams that we shall encounter:

(i) all the strands in the diagram will be oriented;

(ii) all the strands connected to either side of a black box will be oriented alternately in opposite directions (as in §6.2); further, the orientation in a strand will be unaffected in passage through a black box;

(iii) a 'black box' with two sets of n strands will denote an element of $A_0^{0\prime} \cap A_n^0$ or $A_0^{0\prime} \cap A_{n-1}^1$, depending on whether the first strand from the left is oriented upwards or downwards;

(iv) the curves described by the strands will be smooth;

(v) if a strand is the over-strand at some crossing, it will be the over-strand at all crossings it features in; further, as one proceeds along a strand, the parity of the crossings that one comes across will be alternately positive and negative; and finally,

(vi) at a crossing, neither of the strands is allowed to be horizontal.

A *state* on one of these (possibly complicated) diagrams is an assignment of a vertex from $\bigcup_{n,k=0}^1 \pi(A_k^n)$ to each region in the diagram, and of an edge from $\mathcal{G} \cup \mathcal{H} \cup \mathcal{K} \cup \mathcal{L}$ to each segment of each strand in the diagram, such that:

(i) the region on the extreme left is labelled by a vertex from $\pi(A_0^0)$;

(ii) the usual 'homomorphic property' is satisfied;

(iii) at a crossing, the regions surrounding the crossing will be indexed as follows, according to whether the crossing is positive or negative –

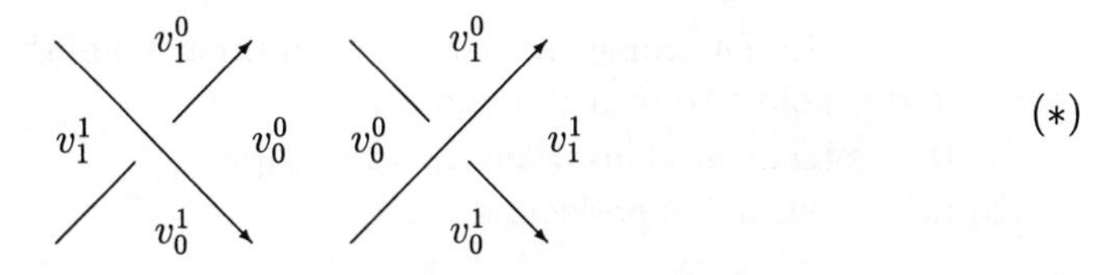

– thus, at a positive (resp., negative) crossing, the region enclosed by the two 'out-arrows' (resp., 'in-arrows') is labelled by a vertex from $\pi(A_0^0)$, and as one proceeds from this vertex in the anticlockwise (resp., clockwise) direction, one will encounter, in order, vertices indexed by $\pi(A_1^0), \pi(A_1^1)$ and $\pi(A_0^1)$;

(iv) the labelling of the regions and strands incident on a black box is consistent with the requirement (determined by the orientation of the strands going into that box) imposed by demanding that that black box is supposed to represent an element of an appropriate relative commutant (see item (iii) in the earlier description of features of our diagrams).

By a 'partial state', we shall mean a state which has been prescribed only on the 'boundary' of the diagram – by which we mean the unbounded regions and unbounded segments of strings. A general diagram is thought of as a function – the 'partition function' – on the set of partial states, as follows: if D is a diagram and if γ is a partial state on the diagram, then the the value Z_D^γ is defined to be the sum, over all states σ which extend γ, of the value of the diagram D on the state γ.

In order to evaluate a diagram on a state, we form the product of all the 'local contributions' (coming from black boxes, local extrema and crossings); the local contribution coming from a black box is determined as before. We now describe how to determine the 'local contribution' coming from (a) a local extremum, and (b) a crossing.

Local extrema: These are, obviously, configurations of either of the following types:

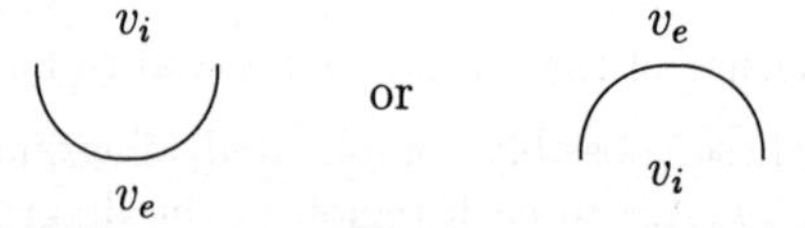

If the 'interior' and 'exterior' regions of such a local extremum are labelled as indicated above, then the 'local contribution' of such an extremum is defined to be

$$\left(\frac{t_{v_i}}{t_{v_e}}\right)^{\frac{1}{2}}.$$

Crossings: To determine the 'local contribution' coming from a crossing, there are two points to bear in mind:

(a) the assignment is invariant under isotopy;

(b) neither strand is horizontal.

We postulate that at a positive or negative crossing of the following form, the associated Boltzmann weight is as indicated.

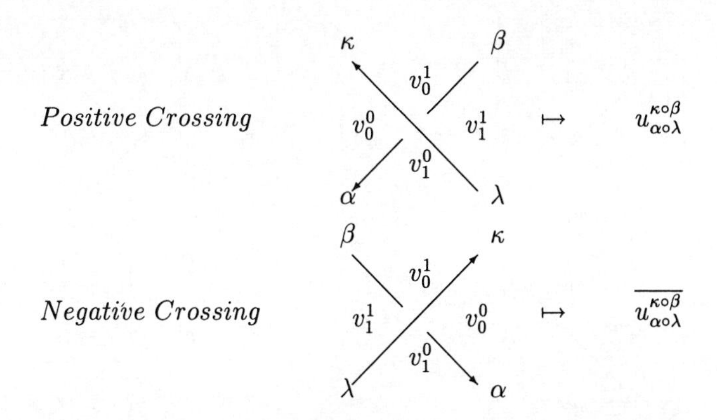

For a crossing which is not in this 'canonical form', use isotopy invariance, as illustrated by the following example:

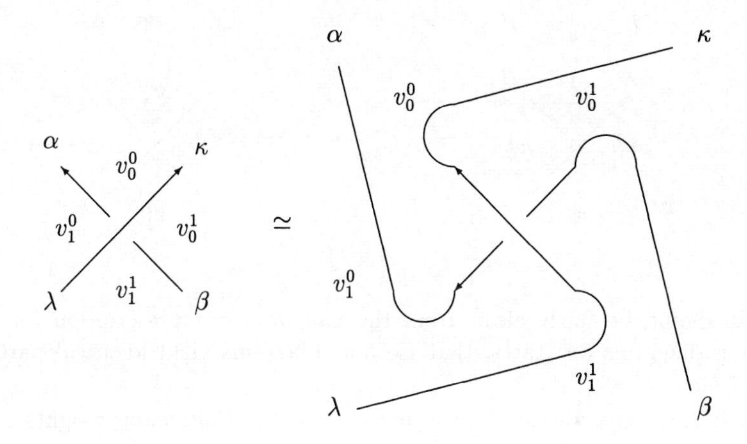

Hence, in accordance with our convention for extrema, the 'local contribution' of the positive crossing given on the left (in 'non-canonical form') is given by

$$u_{\alpha\circ\lambda}^{\kappa\circ\beta} \cdot \left(\frac{\overline{t}_{v_0^0}}{\overline{t}_{v_1^0}}\right)^{\frac{1}{2}} \cdot \left(\frac{\overline{t}_{v_1^1}}{\overline{t}_{v_0^1}}\right)^{\frac{1}{2}}.$$

(Note that the above expression is what, in the notation of §5.5, would have been denoted by $\overline{v_{\tilde{\alpha}\circ\kappa}^{\lambda\circ\tilde{\beta}}}$. We shall use this notation in the immediate sequel, for typographical convenience.)

In an entirely similar fashion, it may be verified that the prescription for assigning Boltzmann weights to the various possible configurations of crossings is as follows:

$$= u^{\kappa\circ\beta}_{\alpha\circ\lambda} =$$

$$= \overline{v^{\lambda\circ\tilde{\beta}}_{\tilde{\alpha}\circ\kappa}} =$$

$$= \overline{u^{\kappa\circ\beta}_{\alpha\circ\lambda}} =$$

$$= v^{\lambda\circ\tilde{\beta}}_{\tilde{\alpha}\circ\kappa} =$$

It should be fairly clear, from the nature of our prescription for evaluating diagrams on states, that isotopic diagrams yield identical partition functions.

Alternatively, we could have just defined the Boltzmann weights for all possible crossings (in all possible 'non-canonical forms') by the preceding prescription and then verified that this was an isotopy-invariant prescription.

In the rest of this section, we give some indications of the sort of advantages that this formalism has over the corresponding formulations with formulae.

(1) To start with, it is a pleasant exercise to check that the matrix u satisfies the biunitarity condition precisely when diagrams related by Reidemeister moves of type II yield identical partition functions.

(2) Next, the inclusion of $A_0^{0\prime} \cap A_n^0$ into $A_0^{0\prime} \cap A_n^1$ is given by the following identification, as can be verified by another pleasant little exercise:

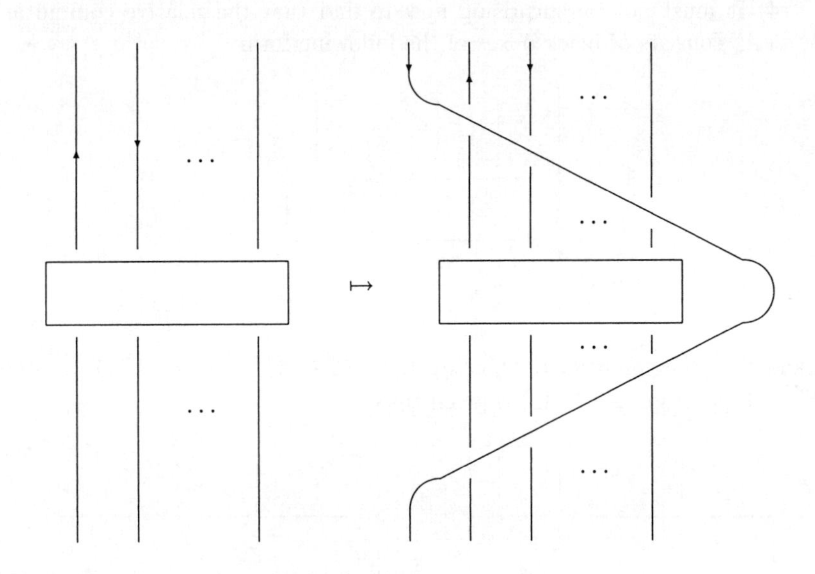

(3) The projection e_n is represented by the following picture –

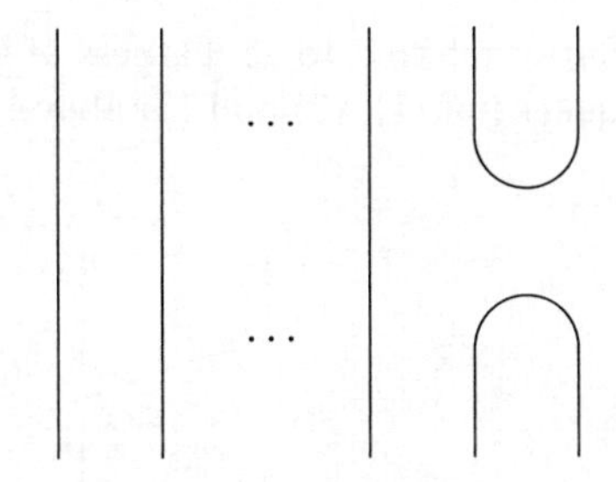

– with the understanding, of course, that if e_n is viewed as an element of $A_0^{0'} \cap A_n^0$ (resp., $A_0^{0'} \cap A_n^1$), then there are n (resp., $n+1$) strands going through the above 'black box'. It follows – from this prescription, and the equation $e_{n+1} x e_{n+1} = (E_{A_{n-1}^k} x) e_n \ \forall x \in A_n^k$ – that the conditional expectation of $A_0^{0'} \cap A_{n+1}^k$ onto $A_0^{0'} \cap A_n^k$ is given thus:

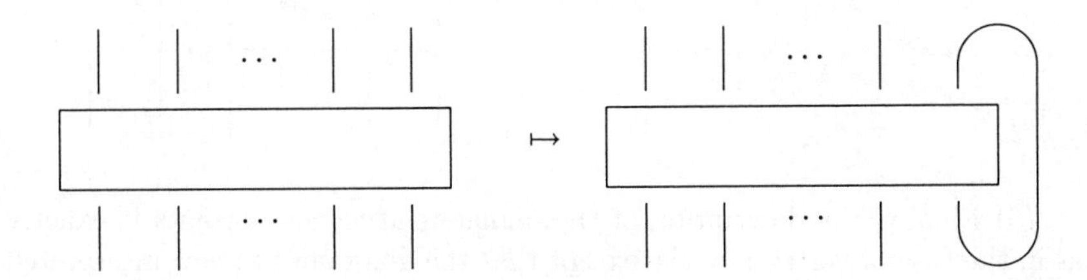

(4) It must not be surprising now to find that the relative commutant $A_1^{k'} \cap A_n^k$ consists of black boxes of the following form –

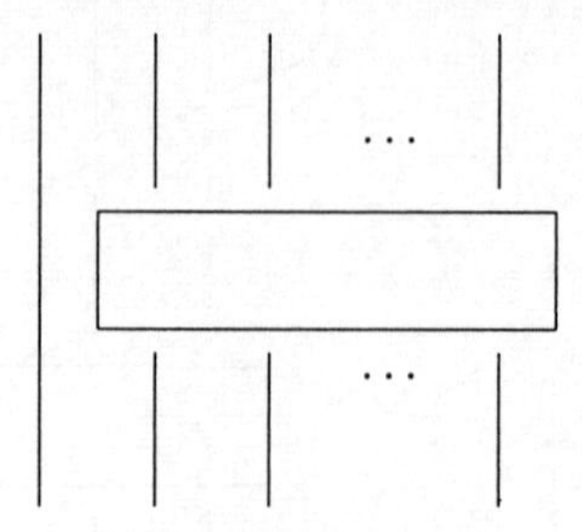

– and that the conditional expectation of $A_0^{0'} \cap A_{n+1}^0$ (resp., $A_0^{0'} \cap A_n^1$) onto $A_1^{0'} \cap A_{n+1}^0$ (resp., $A_0^{1'} \cap A_n^1$) is given thus:

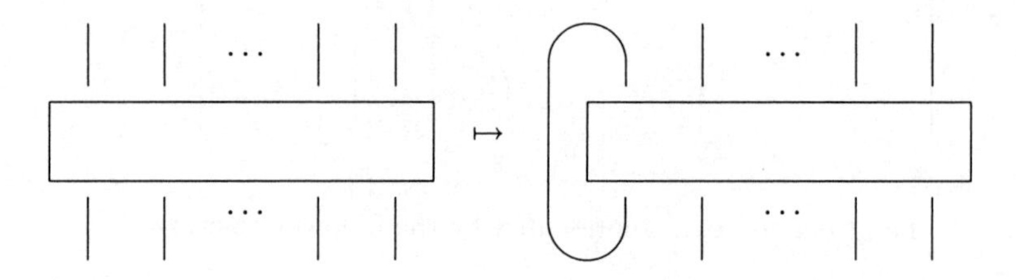

(5) What is customarily referred to as 'flatness of the Jones projections' is an immediate consequence of (1), (2) and (3) above: it is just the asserion that

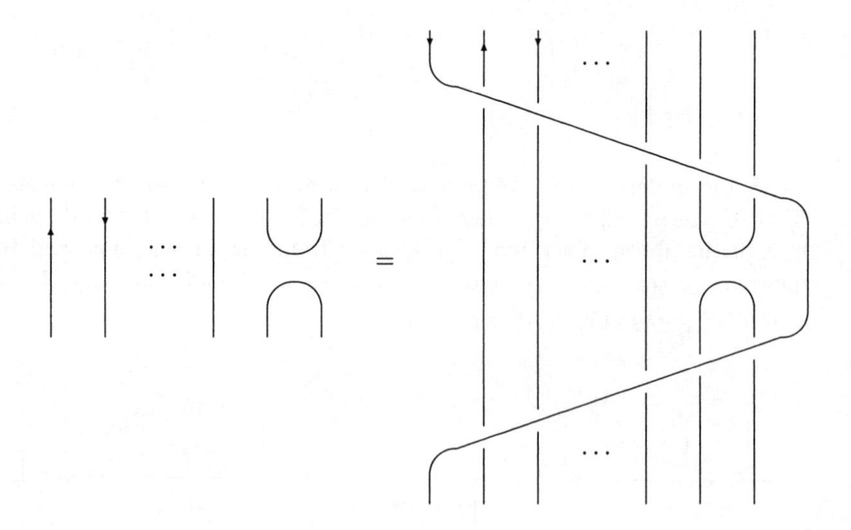

(6) Finally, the description of the higher relative commutants is exactly as in the case of vertex models, except that the diagrams are now interpreted

according to the prescriptions of this section; we state this explicitly as a proposition, whose proof we omit since that is also exactly as for the case of vertex models.

PROPOSITION 6.4.1 *If $n > 0$, then $R_0' \cap R_n$ consists of precisely those $F \in A_0^{0\prime} \cap A_n^0$ for which there exists a $G \in A_0^{1\prime} \cap A_n^1$ such that the following equation holds:*

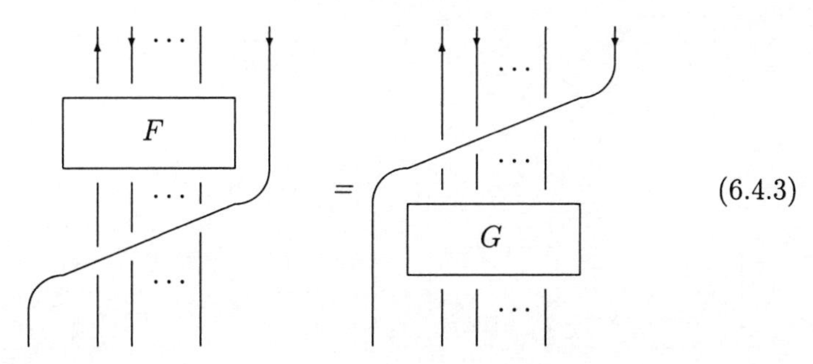

Appendix

A.1 Concrete and abstract von Neumann algebras

We used the word 'concrete' in the opening paragraphs of the first section of this book, to indicate that we were looking at a concrete realisation or representation (as operators on Hilbert space) of a more abstract object. The abstract notion is as follows: suppose M is a C^*-algebra – i.e., a Banach *-algebra, where the involution satisfies $\|x^*x\| = \|x\|^2$ for all x in M; suppose further that M is a dual space as a Banach space – i.e., there exists a Banach space M_* (called the pre-dual of M) such that M is isometrically isomorphic, as a Banach space, to the dual Banach space $(M_*)^*$; let us temporarily call such an M an *'abstract von Neumann algebra'*.

It turns out – cf. [Tak1],Corollary III.3.9 – that the pre-dual of an abstract von Neumann algebra is uniquely determined up to isometric isomorphism; hence it makes sense to define *the σ-weak topology* on M as $\sigma(M, M_*)$, *the* weak* topology on M defined by M_*.

The natural morphisms in the category of von Neumann algebras are *-homomorphisms which are continuous relative to the σ-weak topology (on range as well as domain); such maps are called normal homomorphisms.

The canonical commutative examples of abstract von Neumann algebras turn out to be $L^\infty(X, \mu)$, while the basic non-commutative example is $\mathcal{L}(\mathcal{H})$. (The pre-dual $\mathcal{L}(\mathcal{H})_*$ is the space of trace-class operators on $\mathcal{H}$, endowed with the trace-norm – the duality being given by $\mathcal{L}(\mathcal{H}) \times \mathcal{L}(\mathcal{H})_* \ni (x, \rho) \mapsto \operatorname{tr}(\rho x)$.)

It is not hard to see that if M is an abstract von Neumann algebra, so is N, where N is any σ-weakly closed self-adjoint subalgebra. In particular, a σ-weakly closed self-adjoint subalgebra of $\mathcal{L}(\mathcal{H})$ is an abstract von Neumann algebra.

It follows from the double-commutant theorem that any 'concrete' von Neumann algebra is an 'abstract' von Neumann algebra.

The (artificial) distinction between the notions of abstract and concrete von Neumann algebras may (and shall henceforth) be dispensed with, in view of the following theorem (see [Tak1],Theorem III.3.5): any abstract von Nemann algebra admits a normal *-isomorphism onto a concrete von Neumann algebra.

A.2 Separable pre-duals, Tomita–Takesaki theorem

As was remarked in §1.2, any separable Hilbert space which carries a normal representation of a von Neumann algebra is expressible as a countable direct sum of 'GNS' representations. This fact leads to the following fact.

PROPOSITION A.2.1 *The following conditions on a von Neumann algebra M are equivalent:*

(i) M admits a faithful normal representation on separable Hilbert space;

(ii) M_* is separable.

Proof: (i) $\Rightarrow$ (ii): If $\pi : M \to \mathcal{L}(\mathcal{H})$ is a faithful normal *-representation, it is a fact – cf. [Tak1], Proposition III.3.12 – that $\pi(M)$ is a von Neumann subalgebra of $\mathcal{L}(\mathcal{H})$ and that π is a σ-weak homeomorphism of M onto $\pi(M)$. Hence $M_* \cong \pi(M)_*$, but $\pi(M)_* \cong \mathcal{L}(\mathcal{H})_*/\pi(M)_\perp$, where $\pi(M)_\perp = \{\rho \in \mathcal{L}(\mathcal{H})_* : \langle \pi(x), \rho \rangle = 0 \ \ \forall x \in M\}$, whence $\pi(M)_*$ inherits separability from $\mathcal{L}(\mathcal{H})_*$.

(ii) $\Rightarrow$ (i): In general, if X is a separable Banach space, then ball X^* (the unit ball of the Banach dual space X^*) is a compact metric space (w.r.t. the distance defined by $d(\varphi, \psi) = \sum_{n=1}^{\infty} 2^{-n} |\varphi(x_n) - \psi(x_n)|$ where $\{x_n\}_{n=1}^{\infty}$ is a dense sequence in ball X) and hence separable in the weak*-topology. It follows that $X^* (= \bigcup_n n(\text{ball} X^*))$ is also weak*-separable.

In particular, if M_* is separable, then there exists a sequence $\{x_n\}_{n=1}^{\infty}$ in M which is σ-weakly dense in M. So, if π is a representation on $\mathcal{H}$ with cyclic vector ξ, then $\mathcal{H}$ must be separable since $\{\pi(x_n)\xi\}_{n=1}^{\infty}$ is a countable dense set in $\mathcal{H}$. In particular, if M_* is separable, then $\mathcal{H}_\varphi$ is separable for every φ in M_* (where $(\mathcal{H}_\varphi, \pi_\varphi, \xi_\varphi)$ is the GNS triple for φ).

If $\{\psi_n\}_{n=1}^{\infty}$ is a dense sequence in M_*, clearly $\{\psi_n\}_{n=1}^{\infty}$ separates points in M. Now each ψ_n is expressible – cf. [Sak1], Theorem 1.14.3 – as a linear combination of four normal states on M. Hence there exists a sequence $\{\varphi_n\}_{n=1}^{\infty}$ of normal states on M which separates points in M. Let $(\mathcal{H}_n, \pi_n, \xi_n)$ be the GNS triple associated with φ_n, and let $\mathcal{H} = \bigoplus_n \mathcal{H}_n, \pi = \bigoplus_n \pi_n$. If ξ_n denotes the vector in $\mathcal{H}$ with ξ_n in the n-th co-ordinate and zero in other co-ordinates, it is clear that

(i) $\mathcal{H}$ is separable (by the previous paragraph); and

(ii) $\langle \pi(x)\tilde{\xi}_n, \tilde{\xi}_n \rangle = \varphi_n(x) \ \ \forall x \in M, n = 1, 2, \cdots$.

In particular, $\pi(x) = 0$ implies $\varphi_n(x) = 0 \ \ \forall n$, whence $x = 0$; i.e., π is a faithful normal representation of M on the separable Hilbert space $\mathcal{H}$. □

REMARK A.2.2 *If M satisfies the equivalent conditions of Proposition A.2.1, then M admits a faithful normal state φ. (Reason: if $\mathcal{H}$ is separable, then $\mathcal{L}(\mathcal{H})$ admits a faithful normal state, for instance $\rho = \sum(1/2^n)\langle \cdot, \xi_n \rangle \xi_n$, where $\{\xi_n\}$ is an orthonormal basis for $\mathcal{H}$.)*

Assume for the rest of this section, that M is a von Neumann algebra with separable pre-dual. Thus, by the preceding remark, we may always find a faithful normal state , say φ on M. Let $\mathcal{H} = L^2(M, \varphi)$. It is true, as in the case of finite M, that $\mathcal{H}$ admits a cyclic and separating vector Ω. What is different is that vectors of the form $x\Omega, x \in M$, are, in general, no longer right-bounded. This and other such problems can eventually be overcome, due to the celebrated Tomita–Takesaki theorem. In the following formulation of this theorem, we identify M with its image under the GNS representation π_φ.

THEOREM A.2.3 *Let M, φ be as above. Then the mapping $x\Omega \mapsto x^*\Omega$, defined on the dense subspace $M\Omega$, is a conjugate-linear closable operator S_0. Let $S = J\Delta^{\frac{1}{2}}$ denote the polar decomposition of the closure S of the operator S_0. Then*
(i)$JMJ = M'$; and
(ii)$\mathrm{Ad}_{\Delta^{it}}(M) = M, \ \ \forall t \in \mathbb{R}$. □

One consequence of (i) of this theorem, and Proposition 2.1.2, is that isomorphism classes of separable modules over an arbitrary von Neumann algebra M with separable pre-dual are in bijective correspondence with Murray–von Neumann equivalence classes of projections in $M \otimes \mathcal{L}(\ell^2)$.

A.3 Simplicity of factors

The purpose of this section is to prove the following result.

PROPOSITION A.3.1 *A factor contains no proper weakly closed ideal.*
A finite factor contains no two-sided ideal.

Proof: Suppose I is a two-sided ideal in a factor M. Polar decomposition shows that $x \in I \Leftrightarrow |x| \in I$. This observation, together with the functional calculus, shows that if $I \neq 0$, then there exists a non-zero projection $p \in I$. (Reason: Write 1_r for the indicator function of the set (r, ∞); if $0 \neq x \in I$, pick r sufficiently small to ensure that $p = 1_r(|x|) \neq 0$. Now define the (bounded measurable) function g by

$$g(t) = \begin{cases} \frac{1}{t} & \text{if } t > r, \\ 0 & \text{otherwise,} \end{cases}$$

and note that $tg(t) = 1_r(t) \ \ \forall t \in \mathbb{R}$, whence $|x|g(|x|) = p$.)

Since we can find partial isometries $\{u_i : i \in I\}$ in M such that $1 = \sum_{i\in I} u_i p u_i$ – with I finite if M is finite – the proof of the proposition is complete. □

COROLLARY A.3.2 *A normal homomorphism of a factor is either identically zero or injective.*

A.4 Subgroups and subfactors

We first observe that for an automorphism of a factor, the condition of being free – see Definition 1.4.2 – is equivalent to not being an inner automorphism. (It is obvious that an inner automorphism is not free. Conversely, suppose θ is an automorphism of a factor P and that there exists an element $x \in P$ such that $xy = \theta(y)x, \ \forall y \in P$; this is seen to imply that both x^*x and xx^* belong to the centre of P, and an appeal to polar decomposition suffices to show that either $x = 0$, or θ is inner.)

In this section, we assume that $\alpha : G \to \mathrm{Aut}(P)$ is an outer action of a finite group on a II_1 factor P – i.e., we assume that α is an action such that if $G \ni t \neq 1$, then α_t is not an inner automorphism of P. Since P is a factor, it follows from the last paragraph and Proposition 1.4.4(i) that $P' \cap (P \times_\alpha G) = \mathbb{C}$, and in particular, the crossed product $P_1 = P \times_\alpha G$ is also a II_1 factor. It turns out that P_1 admits a natural action on $L^2(P)$ thus: since the trace on P is unique, it follows easily that there is a unitary representation $t \mapsto u_t$ of G on $L^2(P)$, such that $u_t x\Omega = \alpha_t(x)\Omega$; an easy computation shows that $u_t x u_t^* = \alpha_t(x) \ \ \forall x \in P, t \in G$. This implies that there is a natural homomorphism of $P \times_\alpha G$ onto $(P \cup \{u_t : t \in G\})''$, where, of course, we regard elements of P as left-multiplication operators on $L^2(P)$. Deduce now from Corollary A.3.2 that this homomorphism must be an isomorphism. Hence we assume, in the sequel, that $P_1 = P \cup \{u_t : t \in G\}''$.

In this section, we shall prove the two succeeding propositions and compute the the principal and dual graphs for the inclusion $K \subseteq P_1$, where K is as in Proposition A.4.2 below. (We continue to use the preceding notation in the next two propositions.)

PROPOSITION A.4.1 *If $P_0 = P^G$ is the fixed-point algebra for the G-action, then P_0 is an irreducible subfactor of P such that the result $\langle P, e_{P_0}\rangle$ of the basic construction for the inclusion $P_0 \subseteq P$ coincides with P_1. (Thus, forming the crossed product is 'dual' to taking the fixed-point algebra.)*

PROPOSITION A.4.2 *The passage $H \mapsto K = P \times_{\alpha|_H} H$ establishes a bijective correspondence between subgroups H of G, and *-algebras K satisfying $P \subseteq K \subseteq P_1$.*

Proof of Proposition A.4.1: If J denotes the modular conjugation operator on $L^2(P)$, note that J commutes with each u_t – since automorphisms preserve adjoints – and so

$$JP_1'J = J\langle P, \{u_t\}_{t\in G}\rangle' J = P \cap \{u_t\}'_{t\in G} = P_0.$$

Thus P_0 is indeed a II_1 factor, and P_1 is the result of the basic construction for the inclusion $P_0 \subseteq P$.

Further,

$$J(P_0' \cap P)J = P_1 \cap P' = \mathbb{C},$$

and so P_0 is, indeed, an irreducible subfactor of P, and the proof is complete. □

Before proceeding further, notice that $\{u_t : t \in G\}$ is a basis for P_1/P in the sense of §4.3.

Proof of Proposition A.4.2: A moment's thought shows that, since G is finite, it suffices to prove the following assertion:

Assertion: If $P \subseteq K \subseteq P_1$ is an intermediate *-subalgebra, if $0 \neq x = \sum_{t\in G} a_t u_t \in K, a_t \in P$, and if $S = \{t \in G : a_t (= E_P(xu_t^*)) \neq 0\}$ (denotes the support of x), then there exists $t \in S$ such that $u_t \in K$.

First notice that the assumption $P \subseteq K$ implies that $E_P(Ku_s)$ is a two-sided ideal in P, and must hence be one of the trivial ideals. Hence if x, S are as in the assertion, then $E_P(PxPu_s^*) = P$ for all $s \in S$.

We prove the assertion by induction on the cardinality of S. If $S = \{s\}$ is a singleton, the last observation shows that there exists $a_i, b_i \in P$ such that $E_P(\sum_i a_i x b_i u_s^*) = 1$, which implies, together with the assumption $S = \{s\}$, that $u_s = \sum_i a_i x b_i \in K$.

Suppose then that $|S| > 1$; fix $s \in S$. Argue as above to find $z \in PxP$ such that $E_P(zu_s^*) = 1$. Note that the support S_z of z (as described in the statement of the assertion) is contained in that of x, i.e., $S_z \subseteq S$. Further, $s \in S_z$, by construction. If $S_z = \{s\}$, we are done by the last paragraph. If not, suppose $z = \sum_{k\in S_z} a_k u_k$, with $a_t \neq 0, t \neq s$. Since $\alpha_{ts^{-1}}$ is free, we can find a unitary element $v \in P$ such that $va_t \neq a_t \alpha_{ts^{-1}}(v)$; this means that the element $y = z - vz\alpha_{s^{-1}}(v^*)$ is a non-zero element of K such that $S_y \subseteq (S \setminus \{s\})$, and an appeal to the induction hypothesis completes the proof. □

Assume, in the rest of this section, that $N \subseteq M$ is an inclusion of II_1 factors with $[M : N] = d < \infty$. Once and for all, fix an integer $n \geq d$, set $I = \{1, 2, \cdots, n\}$, fix a basis $\{\lambda_i : i \in I\}$ for M/N, in the sense of §4.3, and consider the (co-finite morphism of M into N given by the) map $\theta : M \to M_I(N)$ defined by $\theta_{ij}(x) = E_N(\lambda_i x \lambda_j^*)$.

Let us write $\mathcal{H}$ for $L^2(M)$, when viewed as an N-M-bimodule. In the language of §4.1, and with the preceding notation, we then have $\mathcal{H} \cong \mathcal{H}_\theta (=$

$M_{1\times n}(L^2(N))\theta(1))$. (In the following, if $\phi : Q \to M_l(P)$ is a co-finite morphism, we shall write $\mathcal{H}_\phi$ to denote the P-Q-bimodule $M_{1\times l}(L^2(P))\phi(1)$. Recall the fact – which we shall use below – that in this case, we have ${}_P\mathcal{L}_Q(\mathcal{H}_\phi) = M_l(P)_{\phi(1)} \cap \phi(Q)'$.) Also, we assume that $M_k, k \geq 1$, are the members of the tower of the basic construction as usual.

LEMMA A.4.3 *With the foregoing notation, for each positive integer k, define $\theta^{(k)} : M \to M_{I^k}(N)$ thus: if $\mathbf{i} = (i_1, \cdots, i_k), \mathbf{j} = (j_1, \cdots, j_k) \in I^k$, then*

$$\begin{aligned} \theta^{(k)}_{\mathbf{i},\mathbf{j}} &= \theta_{i_1 j_1}(\theta_{i_2 j_2}(\cdots \theta_{i_k j_k}(x) \cdots)) \\ &= E_N(\lambda_{i_1} E_N(\lambda_{i_2} E_N(\cdots E_N(\lambda_{i_k} x \lambda^*_{j_k}) \cdots) \lambda^*_{j_2}) \lambda^*_{j_1}). \end{aligned}$$

(i) Then $\theta^{(k)}$ is a co-finite morphism of M into N for each $k \geq 1$, and in fact,

$$ {}_N L^2(M_{k-1})_M \cong \mathcal{H}_{\theta^{(k)}} = M_{1\times n^k}(L^2(N))\theta^{(k)}(1); \tag{A.4.1}$$

consequently, we have an isomorphism of inclusions:

$$\begin{pmatrix} N' \cap M_{2k+1} \\ \cup \\ N' \cap M_{2k} \end{pmatrix} \cong \begin{pmatrix} \theta^{(k+1)}(N)' \cap M_{I^{k+1}}(N)_{\theta^{(k+1)}(1)} \\ \cup \\ \theta^{(k+1)}(M)' \cap M_{I^{k+1}}(N)_{\theta^{(k+1)}(1)} \end{pmatrix}. \tag{A.4.2}$$

(ii) Let $\phi^{(k)} = \theta^{(k)}$, viewed as a map from M into $M_{I^k}(M)$. Then $\phi^{(k)}$ is a co-finite morphism of M into M for each $k \geq 1$, and in fact,

$$ {}_M L^2(M_k)_M \cong \mathcal{H}_{\phi^{(k)}} = M_{1\times n^k}(L^2(M))\theta^{(k)}(1); \tag{A.4.3}$$

consequently, we have an isomorphism of inclusions:

$$\begin{pmatrix} M' \cap M_{2k+1} \\ \cup \\ M' \cap M_{2k} \end{pmatrix} \cong \begin{pmatrix} \theta^{(k)}(N)' \cap M_{I^k}(M)_{\theta^{(k)}(1)} \\ \cup \\ \theta^{(k)}(M)' \cap M_{I^k}(M)_{\theta^{(k)}(1)} \end{pmatrix}. \tag{A.4.4}$$

Proof: First note that equation (A.4.1) implies the co-finiteness of $\theta^{(k)}$ as well as (A.4.2) (in view of Proposition 4.4.1(i) and the parenthetical remark preceding the statement of this lemma); so we only need to prove (i), which we do by induction on k. The case $k = 1$ is valid, by definition; the implication $(A.4.1)_k \Rightarrow (A.4.1)_{k+1}$ is a consequence of Proposition 4.4.1(ii) and the fact (already mentioned at the end of §4.1) that $\mathcal{H}_\theta \otimes \mathcal{H}_\phi \cong \mathcal{H}_{\theta\otimes\phi}$.

The proof of (ii) is similar. □

We shall now apply the preceding lemma to compute the principal and dual graphs for the subgroup-subfactor $N = (P \times H) \subseteq (P \times G) = M$. So assume, for the rest of this section, that P, G, u_t are as in the first part of this section; assume further that H is a subgroup of G. To be specific, suppose $[G : H] = n$, and suppose $G = \coprod_{i=1}^n Hg_i$ is the partition of G into the distinct right-cosets of H. It is clear then that $\{u_{g_i} : 1 \leq i \leq n\}$ is a basis for M/N, and yields the co-finite morphism $\theta : M \to M_n(N)$ given by

$$\theta_{ij}(x) = E_N(u_{g_i} x u_{g_j^{-1}}), \quad \forall x \in M.$$

Notice, in particular, that

$$\theta_{ij}(r) = \delta_{ij}\alpha_{g_i}(r), \quad \forall r \in P, \tag{A.4.5}$$

and

$$\theta_{ij}(u_g) = \left\{ \begin{array}{ll} u_{g_i g g_j^{-1}} & \text{if } g_i g \in H g_j, \\ 0 & \text{otherwise}, \end{array} \right\} \quad \forall g \in G. \tag{A.4.6}$$

In order to discuss $\theta^{(k)}, k \geq 2$, we shall find it convenient to use the following notation: let $I = \{1, 2, \cdots, n\}$; denote the element $(i_1, i_2, \cdots, i_k) \in I^k$ by $\mathbf{i}$; if $\mathbf{i} \in I^k$, define $\sqcap g_{\mathbf{i}} = g_{i_1} g_{i_2} \cdots g_{i_k}$. Also, we shall write $g \mapsto \beta_g^k$ for the action of G on the set I^k defined thus:

$$\beta_g^k(\mathbf{j}) = \mathbf{i} \Leftrightarrow g_{j_l} g_{j_{l+1}} \cdots g_{j_k} g^{-1} \in H g_{i_l} g_{i_{l+1}} \cdots g_{i_k} \text{ for } 1 \leq l \leq k;$$

we shall also, later, write β^k for the associated permutation representation of G on $\mathbb{C}^{n^k}$.

With the preceding notation, the definitions imply that, for $\mathbf{i}, \mathbf{j} \in I^k, k \geq 1$, we have

$$\theta_{\mathbf{ij}}^{(k)}(r) = \delta_{\mathbf{ij}}\alpha_{\sqcap g_{\mathbf{i}}}(r), \quad \forall r \in P, \tag{A.4.7}$$

and

$$\theta_{\mathbf{ij}}^{(k)}(u_g) = \left\{ \begin{array}{ll} u_{(\sqcap g_{\mathbf{i}}) g (\sqcap g_{\mathbf{j}})^{-1}} & \text{if } \mathbf{i} = \beta_g^k(\mathbf{j}), \\ 0 & \text{otherwise}, \end{array} \right\} \quad \forall g \in G. \tag{A.4.8}$$

Now fix $X = ((x_{\mathbf{ij}})) \in M_{I^k}(M)$; the fact that $P' \cap M = \mathbb{C}$ is seen to imply that $X \in \theta^{(k)}(P)'$ if and only if there exist scalars $C_{\mathbf{ij}} \in \mathbb{C}$ such that

$$x_{\mathbf{ij}} = C_{\mathbf{ij}} u_{(\sqcap g_{\mathbf{i}})(\sqcap g_{\mathbf{j}})^{-1}} \quad \forall \mathbf{i}, \mathbf{j} \in I^k. \tag{A.4.9}$$

Another easy computation shows that $X \in \theta^{(k)}(M)'$ if and only if X is given by equation (A.4.9), where the scalars $C_{\mathbf{ij}}$ satisfy the relations

$$C_{\mathbf{ij}} = C_{\beta_g^k(\mathbf{i}), \beta_g^k(\mathbf{j})} \quad \forall g \in G, \mathbf{i}, \mathbf{j} \in I^k. \tag{A.4.10}$$

It follows immediately, from Lemma A.4.3, that we have an isomorphism of inclusions:

$$\left(\begin{array}{c} M' \cap M_{2k+1} \\ \cup \\ M' \cap M_{2k} \end{array} \right) \cong \left(\begin{array}{c} \beta^k|_H(H)' \\ \cup \\ \beta^k(G)' \end{array} \right). \tag{A.4.11}$$

Thus, in the Bratteli diagram for the inclusion $(M' \cap M_{2k}) \subseteq (M' \cap M_{2k+1})$, the central summands of $M' \cap M_{2k}$ (resp., $M' \cap M_{2k+1}$) are indexed by those irreducible representations of G (resp., H) which feature in the representation

β^k (resp., $\beta^k|_H$), and it is clear that the number of bonds joining a vertex indexed by a suitable $\pi \in \hat{G}$ to a vertex indexed by a suitable $\rho \in \hat{H}$ is precisely the multiplicity with which ρ occurs in $\pi|_H$.

Now, suppose $X = ((x_{\mathbf{ij}})) \in M_{I^k}(N)$; since $u_g \in N \Leftrightarrow g \in H$, we find that $X \in \theta^{(k)}(P)'$ if and only if there exist scalars $C_{\mathbf{ij}} \in \mathbb{C}$ such that

$$x_{\mathbf{ij}} = \left\{ \begin{array}{ll} C_{\mathbf{ij}} u_{(\sqcap g_{\mathbf{i}})(\sqcap g_{\mathbf{j}})^{-1}} & \text{if } (\sqcap g_{\mathbf{i}})(\sqcap g_{\mathbf{j}})^{-1} \in H, \\ 0 & \text{otherwise,} \end{array} \right\} \tag{A.4.12}$$

Thus, the relative commutant $\theta^{(k)}(P)' \cap M_{I^k}(N)$ gets identified with the set

$$\mathcal{C}_k = \{((C_{\mathbf{ij}})) \in M_{I^k}(\mathbb{C}) : C_{\mathbf{ij}} = 0 \text{ if } (\sqcap g_{\mathbf{i}})(\sqcap g_{\mathbf{j}})^{-1} \notin H\}.$$

Consider the mapping

$$\bigoplus_{p=1}^{n} M_{I^{k-1}}(\mathbb{C}) \ni \bigoplus_{p=1}^{n} ((C^{(p)}_{\mathbf{i}',\mathbf{j}'})) \mapsto ((C_{\mathbf{ij}})) \in \mathcal{C}_k$$

defined by

$$C_{\mathbf{ij}} = \left\{ \begin{array}{ll} C^{(p)}_{\mathbf{i}_-,\mathbf{j}_-} & \text{if } \sqcap g_{\mathbf{i}}, \sqcap g_{\mathbf{j}} \in H g_p, 1 \leq p \leq n, \\ 0 & \text{if } \sqcap g_{\mathbf{i}}, \sqcap g_{\mathbf{j}} \text{ belong to distinct cosets,} \end{array} \right.$$

where we have used the notation $\mathbf{i}_- = (i_2, \cdots, i_k)$ for $\mathbf{i} \in I^k$. A simple verification shows that this mapping is an isomorphism of *-algebras.

Thus, we find that $(\theta^{(k)}(P)' \cap M_{I^k}(N)) \cong \bigoplus_{p=1}^{n} M_{I^{k-1}}(\mathbb{C})$. If the element $X \in (\theta^{(k)}(P)' \cap M_{I^k}(N))$ corresponds to the element $\bigoplus_{p=1}^{n}((C^{(p)}_{\mathbf{i}',\mathbf{j}'}))$ under this ismorphism, another simple computation shows that X commutes with $\theta^{(k)}(u_g)$ if and only if

$$C^{(\beta^1_g(p))}_{\beta^{k-1}_g(\mathbf{i}),\beta^{k-1}_g(\mathbf{j})} = C^{(p)}_{\mathbf{i},\mathbf{j}}, \quad \forall p \in I, \mathbf{i}, \mathbf{j} \in I^{k-1}. \tag{A.4.13}$$

Let us temporarily write $K = P \times G_0$, where G_0 is a subgroup of G, which we will later choose to be G or H. The preceding analysis shows that $X \in (\theta^{(k)}(K)' \cap M_{I^k}(N))$ if and only if the corresponding $C^{(p)}_{\mathbf{i},\mathbf{j}}$'s satisfy equation (A.4.13) for all $g \in G_0$. It follows readily from this description that, if the set I breaks up into l orbits under the action $\beta^1|_{G_0}$, if $g^{(1)}, \cdots, g^{(l)}$ is a set containing one element from each of these distinct orbits, and if $G_0^{(i)}$ is the isotropy subgroup of G_0 corresponding to the point $g^{(i)}$, then

$$(\theta^{(k)}(K)' \cap M_{I^k}(N)) \cong \bigoplus_{i=1}^{l} \beta^{k-1}(G_0^{(i)})'. \tag{A.4.14}$$

When $G_0 = G$, since the action β^1 of G is clearly transitive, we have $l = 1$, so we may choose $g^{(1)} = 1$, in which case we find that $G_0^{(1)} = H$.

When $G_0 = H$, the orbits under the action $\beta^1|_H$ of H correspond to the double cosets of H in G, and so if l is the number of double cosets, and if

$G = \coprod_{i=1}^{l} Hg^{(i)}H$ is the partition of G into double cosets of H, we find that $G_0^{(i)} = H \cap g^{(i)^{-1}} Hg^{(i)} = H_i$ (say).

It follows, from Lemma A.4.3, equation (A.4.14) and the last two paragraphs, that in the Bratteli diagram for the inclusion $(N' \cap M_{2k}) \subseteq (N' \cap M_{2k+1})$, the central summands of $(N' \cap M_{2k})$ (resp., $(N' \cap M_{2k+1})$) are indexed by those irreducible representations of H (resp., H_i for any i) which occur in $\beta^{k-1}|_H$ (resp., $\beta^{k-1}|_{H_i}$), and that the vertex labelled by a suitable $\pi \in \hat{H}$ is connected to the vertex labelled by a suitable $\rho \in \hat{H}_i, 1 \leq i \leq l$, by m bonds, if m is the multiplicity with which ρ occurs in $\pi|_{H_i}$.

A moment's thought about the nature of the principal and dual graphs should convince the reader that we have proved the following.

PROPOSITION A.4.4 *Let H be a subgroup of finite index in a discrete group G. Suppose G acts as outer automorphisms on the II_1 factor P. Then the principal graph $\mathcal{G}$ and the dual graph $\mathcal{H}$ for the inclusion $N = P \times H \subseteq P \times G = M$ have the following descriptions.*

Let $G = \coprod_{i=1}^{l} Hg^{(i)}H$ be the partition of G into double cosets of H, and let $H_i = H \cap g^{(i)^{-1}} Hg^{(i)}$. First define a bipartite graph $\tilde{\mathcal{G}}$ as follows: let $\tilde{\mathcal{G}}^{(0)} = (\coprod_{i=1}^{l}(\hat{H}_i \times \{i\})) \times \{0\}, \tilde{\mathcal{G}}^{(1)} = \hat{H} \times \{1\}$; join the even vertex $((\rho, i), 0)$ to the odd vertex $(\pi, 1)$ by m bonds, if m is the multiplicity with which ρ occurs in $\pi|_{H_i}$. Then $\mathcal{G}$ is the connected component in $\tilde{\mathcal{G}}$ which contains the odd vertex $(1, 1)$ which is indexed by the trivial representation of H.

Define the bipartite graph $\tilde{\mathcal{H}}$ as follows: let $\mathcal{H}^{(0)} = \hat{G} \times \{0\}, \mathcal{H}^{(1)} = \hat{H} \times \{1\}$; connect the even vertex $(\pi, 0)$ to the odd vertex $(\rho, 1)$ by as many bonds as the multiplicity with which ρ occurs in $\pi|_H$. Then $\mathcal{H}$ is the connected component in $\tilde{\mathcal{H}}$ which contains the even vertex $(1, 0)$ indexed by the trivial representation of G. □

A.5 From subfactors to knots

In this section, we briefly sketch the manner in which the initial contact between von Neumann algebras and knot theory was made; to be precise, we outline the construction and some basic properties of what has come to be known as the *one-variable Jones polynomial invariant of links.*

The starting point is *Artin's n-strand braid group*, which we briefly describe. Fix a positive integer n – which should be at least 2 for anything interesting to happen. Consider two horizontal rods with n hooks on each of them, and suppose n strands, say of rope, are tied with one end to each of the rods, in such a way that no hook has more than one strand tied to it. In order to avoid pathologies, we assume that the two rods are placed horizontally with one vertically above the other, and that the passage from the top rod to the bottom is not allowed to 'double back', meaning that at any intermediate height, there is exactly one point of each of the n strands.

An n-strand braid is an equivalence class of such arrangements, where two such arrangements are considered to be equivalent if it is possible to continuously deform the one to the other. We shall think of the strands of a braid as being oriented from the top to the bottom. Consistent with this convention, we define the product of two n-strand braids by concatenation, as follows:

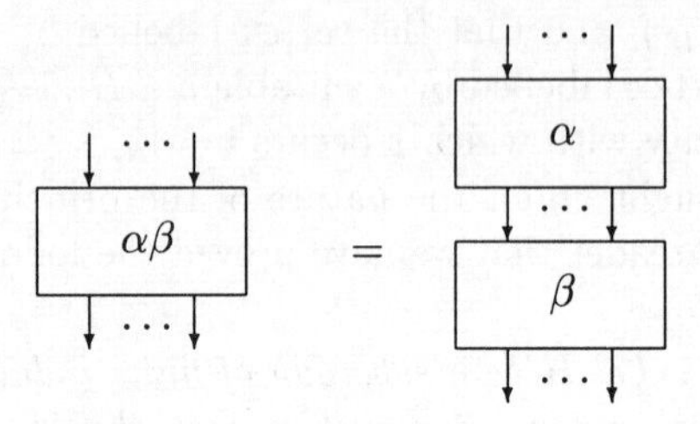

It is fairly painless to verify that this definition endows the set B_n of n-strand braids with the structure of a group. (The inverse of a braid is given by the braid obtained by reflecting the given braid in a mirror placed on the horizontal plane through the bottom rod.) We shall only consider *tame* braids, by which we mean one which admits only a finite number of crossings. It then follows from our definition that B_n is generated, as a group, by the set $\{\sigma_1, \cdots, \sigma_{n-1}\}$, where σ_i is a braid with only one crossing, which is between the $(i-1)$-th and i-th strands and which is positive according to the convention adopted in §6.1. Thus, for instance, when $n = 2$, the generator and its inverse are given as follows:

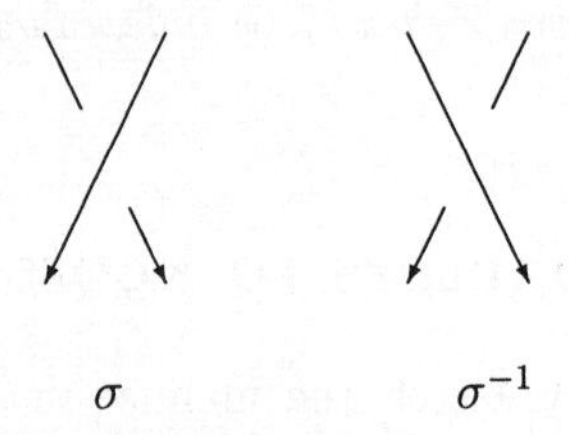

σ σ^{-1}

A result due to Artin establishes the precise relations between these generators. (The reader will find it instructive to draw some pictures to convince herself that these relations are indeed satisfied.)

THEOREM A.5.1 *The braid group has the following presentation (in terms of generators and relations):*

$$B_n = \left\langle \sigma_1, \cdots, \sigma_{n-1} : \begin{array}{ll} (b1) & \sigma_i\sigma_j = \sigma_j\sigma_i \text{ if } |i-j| > 1, \\ (b2) & \sigma_i\sigma_{i+1}\sigma_i = \sigma_{i+1}\sigma_i\sigma_{i+1} \end{array} \right\rangle. \qquad \square$$

In other words, the theorem says that if $g_1, \cdots, g_{n-1}$ are arbitrary elements in a group G, then a necessary and sufficient condition for the existence of a homomorphism $\phi : B_n \to G$ such that $\phi(\sigma_i) = g_i \ \forall i$ is that the g_i's satisfy the so-called braid relations $(b1)$ and $(b2)$, and in this case such a ϕ is unique. Two consequences are worth singling out.

REMARK A.5.2 *(1) There exists a unique epimorphism $\pi : B_n \to S_n$ such that $\pi(\sigma_i)$ is the transposition $(i, i+1)$. (It should be clear that in the language of rods and strands, the braid α is such that the strand which ends at the i-th hook on the bottom rod starts at the $(\pi(\alpha))(i)$-th strand of the top rod.)*

(2) There exists a unique homomorphism $\psi_n : B_n \to B_{n+1}$ such that $\psi_n(\sigma_i^{(n)}) = \sigma_i^{(n+1)}, 1 \leq i < n$ – where we have used the notation $\sigma_i^{(k)}$ to denote the i-th generator of B_k. We shall, in the sequel, write

$$\alpha^{(n+1)} = \psi_n(\alpha^{(n)}), \quad \forall \alpha^{(n)} \in B_n. \tag{A.5.1}$$

Alternatively, we can see that

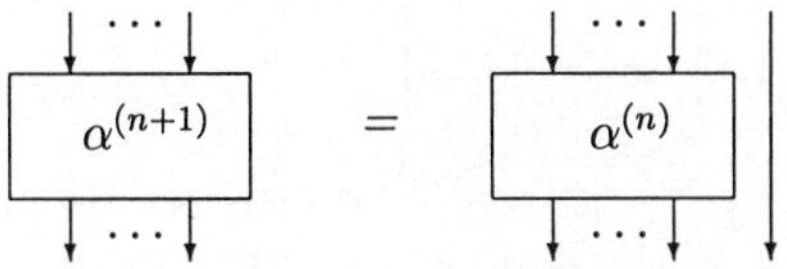

$$\boxed{\alpha^{(n+1)}} \quad = \quad \boxed{\alpha^{(n)}}\ \Big|$$

(3) If we set $\sigma = \sigma_1 \sigma_2 \cdots \sigma_{n-1} \in B_n$, it follows immediately from the braid relations that $\sigma \sigma_i = \sigma_{i+1} \sigma$ for $1 \leq i < n-1$; in other words, the generators σ_i are pairwise conjugate in B_n.

In view of the striking similarity between the braid relations and the relations satisfied by the e_n's, it is natural to try to use the latter to obtain a representation of the former. The simplest way to obtain an invertible element from a projection is to form a (generic) linear combination of the projection and the identity. In view of Remark A.5.2(3), we wish therefore to set

$$g_i = C\{(q+1)e_i - 1\}, 1 \leq i < n, \tag{A.5.2}$$

where C and q are non-zero scalars. Recall that the e_i's come with a parameter τ; an easy computation shows that the g_i's as defined above satisfy the braid relations precisely when the parameters τ and q are related by the equation

$$\tau^{-1} = q + q^{-1} + 2. \tag{A.5.3}$$

Notice that

$$\tau^{-1} = 4\cosh^2 z \Leftrightarrow q = \exp(\pm 2z).$$

Hence, as the parameter q varies over the set $\{\exp(\frac{2\pi\sqrt{-1}}{n}) : n = 3, 4, \cdots\} \cup (0, \infty)$, the parameter τ ranges over all possible index-values of subfactors. Further, the values of q for which the g_i's (with the normalisation $C = 1$) afford a unitary representation of the braid group are precisely the roots of unity.

For convenience of reference, we paraphrase the foregoing remarks in the following proposition.

PROPOSITION A.5.3 *Let $q \in \{\exp(\frac{2\pi\sqrt{-1}}{n}) : n = 3, 4, \cdots\} \cup (0, \infty)$, let τ be defined by equation (A.5.3), and let $\{e_n\}_{n=1}^{\infty}$ be the sequence of projections associated to this τ. Then, for any $C \neq 0$, there exits a unique homomorphism π_n of B_n into the group of units of R such that*

$$\pi_n(\sigma_i^{(n)}) = C\{(q+1)e_i - 1\} \quad \text{for } 1 \leq i < n. \qquad \square$$

The *closure* $\hat{\alpha}$ of a braid $\alpha \in B_n$ is defined as follows:

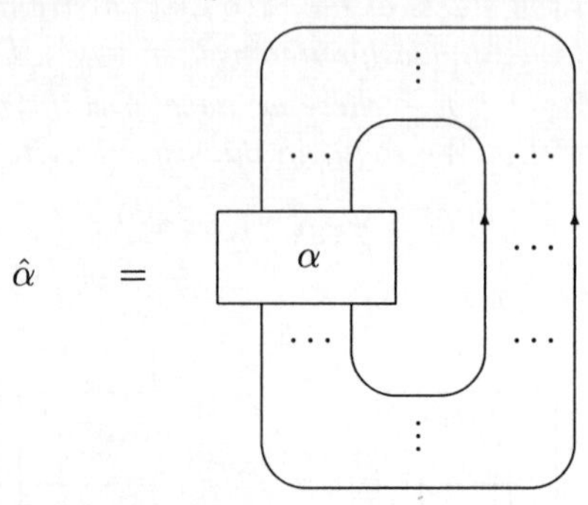

It should be clear that the closure of a braid is an oriented link. (Recall that a *link* is a homeomorphic image of a disjoint union of circles, and that a link is said to be oriented if an orientation has been specified in each component. Alternatively, an oriented link is a compact 1-manifold without boundary, with a distinguished orientation. A *knot* is a link with exactly one component – i.e., a knot is just a homeomorphic image of the circle.) A moment's thought should convince the reader that the number of components of $\hat{\alpha}$ is precisely the number of disjoint cycles in the cycle-decomposition of the permutation $\pi(\alpha)$ – see Remark A.5.2(1).

Two links are considered to be equivalent, or the same, if the one can be continuously deformed to the other. Precisely, this means that there is a continuous map $F : \mathbb{R}^3 \times [0,1] \to \mathbb{R}^3$ such that if we write $F(x,t) = f_t(x)$, then $f_0 = \mathrm{id}_{\mathbb{R}^3}$, each f_t is a homeomorphism of $\mathbb{R}^3$ onto itself, and f_1 maps the first link onto the second. (We only consider links in $\mathbb{R}^3$ here.) Two oriented links are said to be equivalent if they are equivalent as above in such a way that f_1 preserves the orientations. We shall use the symbol $\mathcal{L}$ to denote the class of oriented links (in $\mathbb{R}^3$).

Recall that a link is said to be *tame* if it is equivalent to one which is a smoothly embedded submanifold of $\mathbb{R}^3$.

THEOREM A.5.4 (Alexander) *Every tame link (in $\mathbb{R}^3$) is equivalent to the closure of some braid (on a possibly large number of strands).*

The final ingredient for us to make the connection between subfactors and links is a result due to Markov which explains precisely how two different braids (on possibly different numbers of strands) can have equivalent link closures. In order to describe this result, some terminology would help.

DEFINITION A.5.5 *Let $\alpha^{(n)} \in B_n, \beta^{(m)} \in B_m$. The braids $\alpha^{(n)}$ and $\beta^{(m)}$ are said to be related by a Markov move of (a) type I, if $n = m$ and if $\alpha^{(m)}$ and $\beta^{(m)}$ belong to the same conjugacy class in B_n; and (b) type II, if either (i) $m = n+1$ and $\beta^{(n+1)} = \alpha^{(n+1)}(\sigma_n^{(n+1)})^{\pm 1}$, or (ii) $m = n-1$ and $\beta^{(n)} = \alpha^{(n)}(\sigma_{n-1}^{(n)})^{\pm 1}$.*

A couple of diagrams should convince the reader of the sufficiency of the following condition.

THEOREM A.5.6 (Markov) *In order that two braids $\alpha^{(n)} \in B_n$ and $\beta^{(m)} \in B_m$ have equivalent link-closures, it is necessary and sufficient that there exist braids $\alpha^{(n)} = \alpha_0, \alpha_1, \cdots, \alpha_k = \beta^{(m)}$ such that α_i and α_{i+1} are related by a Markov move (of either type), for $0 \leq i < k$.*

An *invariant of oriented links* (which takes values in some set, say S) is an assignment $\mathcal{L} \ni L \mapsto P_L \in S$ with the property that $L \simeq L' \Rightarrow P_L = P_{L'}$. An immediate consequence of Theorems A.5.4 and A.5.6 is that in order to define an invariant $L \mapsto P_L$ of tame oriented links taking values in S, it is necessary and sufficient to find functions $P_n : B_n \to S, n \geq 2$ such that

$$P_n \text{ is a class function on } B_n, \quad \forall n, \tag{A.5.4}$$

and

$$P_{n+1}(\alpha^{(n+1)}(\sigma_n^{(n+1)})^{\pm 1}) = P_n(\alpha^{(n)}), \quad \forall \alpha^{(n)} \in B_n \quad \forall n; \tag{A.5.5}$$

when this happens, we have

$$P_{\widehat{\alpha^{(n)}}} = P_n(\alpha^{(n)}), \quad \forall \alpha^{(n)} \in B_n \quad \forall n \geq 2.$$

Since class functions are usually obtained by taking the trace of a representation, it is natural to seek a link invariant by considering the functions

$$Q_n(\alpha) = \mathrm{tr}_R(\pi_n(\alpha)),$$

where π_n is as in Proposition A.5.3. (Thus the condition (A.5.4) is automatically satisfied because of the trace.)

As for the condition (A.5.5), note, to start with, that $g_i^{-1} = C^{-1}\{(q^{-1} + 1)e_i - 1\}$, and hence $g_n^{\pm 1}$ satisfies the Markov property with respect to the algebra generated by $\pi_n(B_n)$. Hence, if there is any hope of the Q_n's satisfying (A.5.5), it must at least be the case that $\mathrm{tr}\, g_n = \mathrm{tr}\, g_n^{-1}$; a little algebra shows that this happens precisely when we make the choice $C = q^{\frac{1}{2}}$, in which case, we find that

$$\mathrm{tr}\, g_n^{\pm 1} = -(q^{\frac{1}{2}} + q^{-\frac{1}{2}})^{-1},$$

and consequently, that

$$Q_{n+1}(\alpha^{(n+1)}(\sigma_n^{(n+1)})^{\pm 1}) = \{-(q^{\frac{1}{2}} + q^{-\frac{1}{2}})^{-1}\} Q_n(\alpha^{(n)})$$

for any $\alpha^{(n)} \in B_n$.

A moment's thought shows that if we define

$$P_n(\alpha) = \{-(q^{\frac{1}{2}} + q^{-\frac{1}{2}})\}^{n-1} Q_n(\alpha), \alpha \in B_n,$$

then the P_n's do satisfy both the conditions (A.5.4) and (A.5.5). We have thus proved the following:

THEOREM A.5.7 *Let q, τ, π_n be as in Proposition A.5.3, where we assume that $C = q^{\frac{1}{2}}$; then there exists a complex-valued invariant of oriented links, which we shall denote by $\mathcal{L} \ni L \mapsto V_L(q)$, such that, if $\alpha \in B_n$, then*

$$V_{\widehat{\alpha}}(q) = \{-(q^{\frac{1}{2}} + q^{-\frac{1}{2}})\}^{n-1} \mathrm{tr}\, \pi_n(\alpha). \qquad \square$$

Most properties of this invariant are consequences of the fact that it satisfies the so-called *skein relations.* To see what these are, it will be convenient to use the point of view of *link diagrams.* The fact is that the image of a tame link in $\mathbb{R}^3$ under the projection onto a generic plane in $\mathbb{R}^3$ will have only double points – meaning that the inverse image of a point in the plane will meet the given link in at most two points. In order to fully recapture the link from the projection, it is necessary to indicate, at each crossing, which of the two strands goes 'over' the other. For instance, a diagram representing the so-called right-handed trefoil knot (with an orientation indicated) is

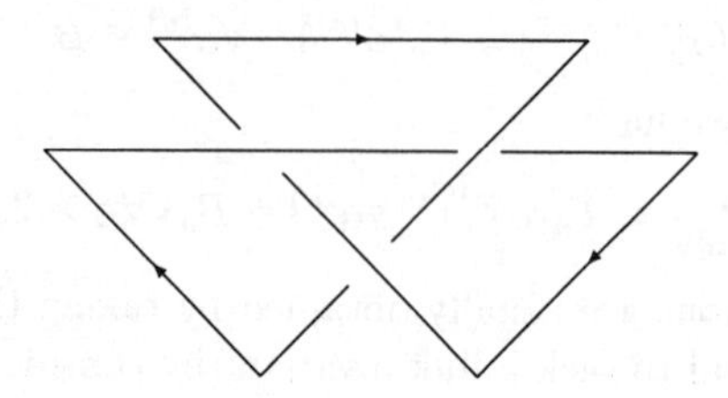

We shall henceforth identify the class $\mathcal{L}$ with the class of all oriented (tame) link diagrams. We shall say that three link diagrams L_+, L_-, L_0 are *skein-related* if they are identical except at one crossing, where they have the following form:

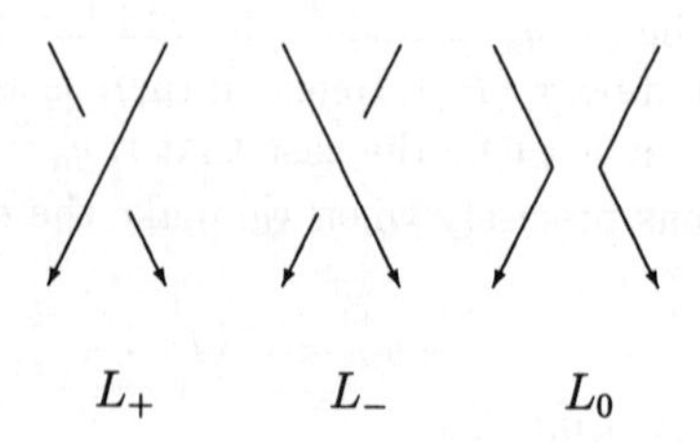

L_+ $\qquad L_-$ $\qquad L_0$

Before stating the next result, we recall that the *unlink U_c with c components* is nothing but $\widehat{1_c}$, where 1_c denotes the identity element of B_c.

PROPOSITION A.5.8 *The invariant defined in Theorem A.5.7 satisfies the following relations:*
(i) $V_{U_c}(q) = \{-(q^{\frac{1}{2}} + q^{-\frac{1}{2}})\}^{n-1}$;
(ii) if L_+, L_- *and* L_0 *are skein-related as above, then*

$$q^{-1}V_{L_+}(q) - qV_{L_-}(q) = (q^{\frac{1}{2}} - q^{-\frac{1}{2}})V_{L_0}(q).$$

Proof: Assertion (i) is an immediate consequence of the definitions.
(ii) Begin by observing that

$$g_i = q^{\frac{3}{2}}e_i - q^{\frac{1}{2}}(1 - e_i),$$

and hence g_i satisfies the quadratic relation

$$(g_i - q^{\frac{3}{2}})(g_i + q^{\frac{1}{2}}) = 0,$$

which may be re-written as

$$q^{-1}g_i - qg_i^{-1} = (q^{\frac{1}{2}} - q^{-\frac{1}{2}})1.$$

The desired conclusion is a consequence of the definition of V_L, properties of the trace, and the fairly obvious observation that the assumed skein-relation between L_+, L_- and L_0 amounts to the existence of a positive integer n, elements α and β of B_n and an integer $1 \leq i < n$ such that – thinking of these link diagrams as links – we have

$$L_+ = \widehat{\alpha\sigma_i^{(n)}\beta}, \quad L_- = \widehat{\alpha(\sigma_i^{(n)})^{-1}\beta}, \quad L_0 = \widehat{\alpha\beta}. \qquad \square$$

The following elementary lemma will be very useful in deducing properties of the invariant V_L from Proposition A.5.8.

LEMMA A.5.9 *Let* L *be an oriented link diagram; there exists a subset of the set of crossings in* L *such that, if we change all these crossings (from an over- to an under-crossing and vice versa), the resulting diagram represents an unknot with the same number of components as* L.

Proof: Arbitrarily label the distinct components of the diagram $1, 2, \cdots, c$, and fix a reference point on each component, which is not a double point. Given any crossing, change it if (and only if) one of the following things happens: either (a) the crossing involves two different components, and the component with the larger label crosses over the one with the smaller label; or (b) the crossing involves only one component, and in travelling along that component from the chosen reference point in the direction specified by the orientation, the first time you come to the crossing, you find yourself going along the over-strand of the crossing.

A moment's thought should convince the reader that this algorithm yields a proof of the lemma.

For instance, if L is the link diagram given earlier to depict the right-handed trefoil, our algorithm would yield the following diagram (with the reference point as indicated):

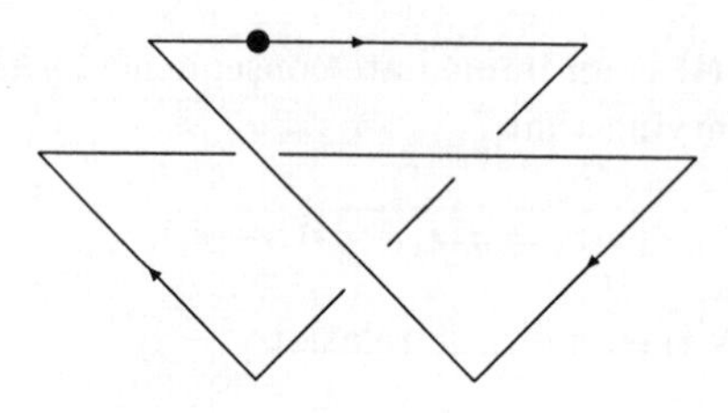

□

Given a link diagram L, define its 'knottiness' to be the ordered pair (n, k), where n denotes the number of crossings in the diagram and k denotes the minimum of the cardinalities of subsets of the set of crossings which satisfy Lemma A.5.9; we shall say that a link diagram L_2 is 'more knotty' than a diagram L_1 if the knottiness (n_1, k_1) of L_1 precedes the knottiness (n_2, k_2) of L_2 in the lexicographic ordering – i.e., if either $n_1 < n_2$, or $n_1 = n_2$, $k_1 < k_2$.

With respect to this 'ordering', the least knotty diagrams represent unlinks, and the preceding lemma has the following pleasing consequence: given any link diagram which does not represent an unlink, there is a triple (L_+, L_-, L_0) of skein-related diagrams such that two things hold: (a) L is either L_+ or L_-, and (b) L is more knotty than the other two diagrams in $\{L_+, L_-, L_0\}$.

Since the set $\mathbb{Z}_+ \times \mathbb{Z}_+$ is well-ordered with respect to the lexicographic order, the preceding considerations allow us to prove facts about V_L using a 'knotty induction'. By such a process, it is easy (and amusing) to prove the following facts. (The strategy of proof is: first prove it for unlinks; then assume the result for L_0 and L_+ (resp., L_-) and use Proposition A.5.8(ii) to deduce the result for L_- (resp., L_+).)

PROPOSITION A.5.10 *(1) $V_L(q)$ is a Laurent polynomial in $q^{\frac{1}{2}}$; more precisely, if L has an odd number of components, then $V_L(q)$ is a Laurent polynomial in q, while if L has an even number of components, then $V_L(q)$ is $q^{\frac{1}{2}}$ times a Laurent polynomial in q.*

(2) If $\tilde{L}$ denotes the mirror-reflection of L, then

$$V_{\tilde{L}}(q) = V_L(q^{-1}).$$

(3) Properties (i) and (ii) of Proposition A.5.8 determine the invariant V_L uniquely (via the process of 'knotty induction' discussed earlier). □

We close by illustrating (3) above with the already mentioned example of the right-handed trefoil.

$$L_+ = T_+ = \textit{right-handed trefoil}$$

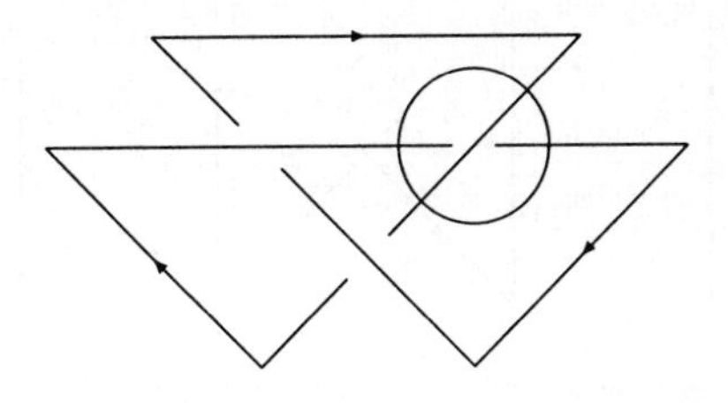

$$L_- = U_1 = \textit{unknot}$$

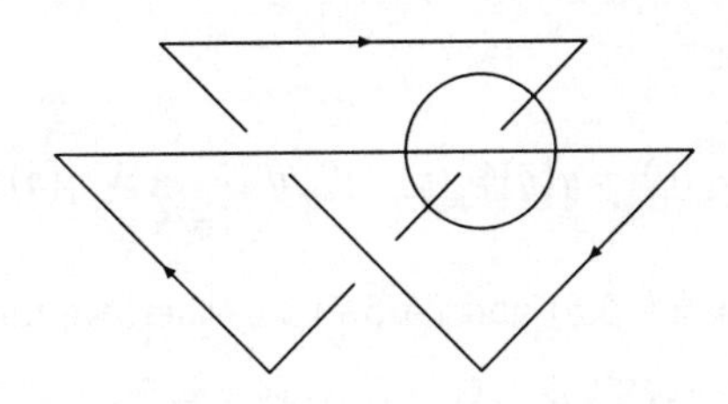

$$L_0 = H_+ = \textit{Hopf link}$$

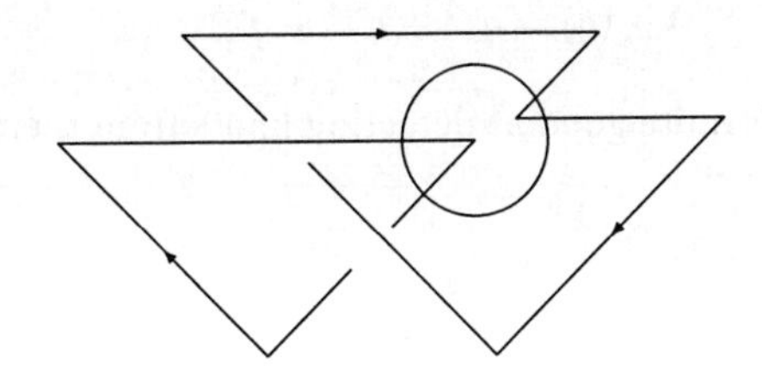

It follows that

$$V_{T_+}(q) = q\{qV_{U_1}(q) + (\sqrt{q} - \frac{1}{\sqrt{q}})V_{H_+}(q)\}. \qquad \text{(A.5.6)}$$

Next, in order to determine V_{H_+}, notice the following skein-related triple of links:

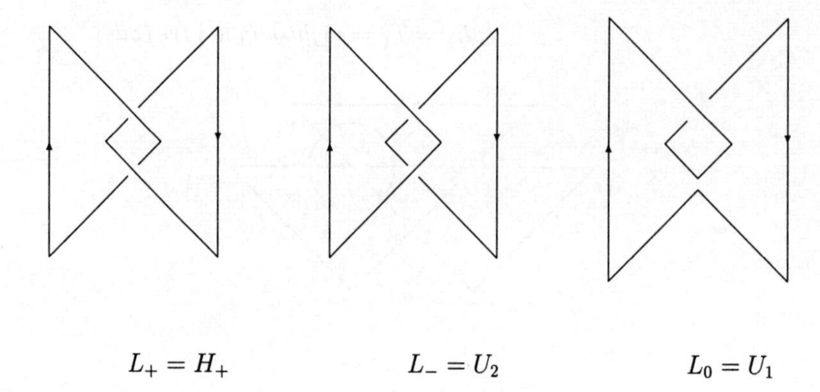

This implies that

$$V_{H_+}(q) = q\{qV_{U_2}(q) + (\sqrt{q} - \frac{1}{\sqrt{q}})V_{U_1}(q)\}. \tag{A.5.7}$$

Putting equations (A.5.6) and (A.5.7) together, we find that

$$V_{T_+}(q) = q + q^3 - q^4,$$

which implies that if $T_- = \tilde{T_+}$ – in the notation of Proposition A.5.10 – so that T_- denotes the so-called *left-handed trefoil* – then

$$V_{T_-}(q) = q^{-1} + q^{-3} - q^{-4}.$$

The invariant V_L is quite good at detecting knots from their mirror-images in this fashion.

Bibliography

[Ake] P. Akemann, *On a class of endomorphisms of the hyperfinite II_1 factor*, preprint.

[Alex] J. Alexander, *A lemma on systems of knotted curves,* Proc. Nat. Acad., 9, (1923), 93–95.

[Art] E. Artin, *Theory of braids,* Ann. Math., 48, (1947), 101–126.

[Aub] P.-L. Aubert, *Deux actions de $SL(2,\mathbb{Z})$*, L'Enseignement Math., 29, (1981), 39–46.

[BHJ] Roland Bacher, Pierre de la Harpe and V.F.R. Jones, *Carres commutatifs et invariants de structures combinatoires,* Comptes Rendus Acad. Sci. Paris, Série 1, 1049–1054, 1995.

[B-N] J. Bion-Nadal, Short Communications, ICM-90.

[Bir] J. Birman, *Braids, links and mapping class groups,* Princeton Univ. Press, 1976.

[Bra] O. Bratteli, *Inductive limits of finite-dimensional C^*-algebras*, Trans. Amer. Math. Soc., 171, (1972), 195–234.

[Ch] E. Christensen, *Subalgebras of a finite algebra*, Math. Ann., 243, (1979), 17–29.

[Con1] A. Connes, *Classification of injective factors*, Ann. Math., 104, (1976), 73–115.

[Con2] A. Connes, *Notes on correspondences*, preprint, 1980.

[Gant] F.R. Gantmacher, *Theory of Matrices, Vol. II,* Chelsea Pub., 1959.

[GN] I.M. Gelfand and M.A. Naimark, *On the embedding of normed rings into the ring of operators in Hilbert space*, Mat. Sb., 12, (1943), 197–213.

[Goo] K.R. Goodearl, *Notes on real and complex C^*-algebras*, Shiva Math. Ser., 5, 1982.

[GHJ] F. Goodman, P. de la Harpe and V.F.R. Jones, *Coxeter graphs and towers of algebras,* MSRI Publ., 14, Springer, New York, 1989.

[Haa] U. Haagerup, *Connes' bicentraliser problem and the uniqueness of the injective factor of type III_1*, Acta Math., 158, (1987), 95–148.

[HS] U. Haagerup and J. Schou, *Some new subfactors of the hyperfinite II_1 factor*, preprint, 1989.

[Had] J. Hadamard, *Resolution d'une question relative aux déterminants*, Bull, des Sci. Math., 17, (1983), 240–246.

[Iz1] M. Izumi, *Application of fusion rules to classification of subfactors*, Publ. RIMS Kyoto Univ., 27, (1991), 953–994.

[Iz2] M. Izumi, *On flatness of the Coxeter graph E_8*, Pac. J. Math., to appear.

[JNM] F. Jaeger, M. Matsumoto and N. Nomura, *Association schemes and spin models*, preprint.

[Jon1] V.F.R. Jones, *Index for subfactors,* Invent. Math.,71,(1983), 1–25.

[Jon2] V.F.R. Jones, *Hecke algebra representations of braid groups and link polynomials*, Ann. Math., 126, (1987), 335–388.

[Jon3] V.F.R. Jones, *On knot invariants related to some statistical mechanical models,* Pacific J. Math., 137, (1989), 311–334.

[Jon4] V.F.R. Jones, *A polynomial invariant for knots via von Neumann algebras*, Bull. Amer. Math. Soc., 12, (1985), 103–111.

[Jon5] V.F.R. Jones, *Notes on subfactors and statistical mechanics*, pp. 1–25, Braid group, knot theory and statistical mechanics, ed. C.N. Yang and M.-L. Ge, World Scientific, 1989.

[Jon6] V.F.R. Jones, *On a family of almost commuting endomorphisms*, J. Functional Anal., 119, (1994), 84–90.

[Kal] R.R. Kallman, *A generalisation of free action*, Duke Math. J., 36, (1969), 781–789.

[Kaw] Y. Kawahigashi, *On flatness of Ocneanu's connection on the Dynkin diagrams, and Classification of Subfactors*, preprint.

[KY] H. Kosaki and S. Yamagami, *Irreducible bimodules associated with crossed-product algebras*, Internat. J. Math., 3, (1992), 661–676.

[KS] Uma Krishnan and V.S. Sunder, *On biunitary permutation matrices and some subfactors of index 9*, Trans. of the Amer. Math. Soc., 348 (1996), 4691-4736.

[KSV] Uma Krishnan, V.S. Sunder and Cherian Varughese, *On some subfactors of integer index arising from vertex models*, J. Functional Anal., 140, (1996), 449–471.

[Mark] A. Markov, *Über die freie Äquivalenz geschlossener Zöpfe*, Math. Sb. 1, (1935), 73–78.

[MvN1] F. Murray and J. von Neumann, *On rings of operators*, Ann. Math., 37, (1936), 116–229.

[MvN2] F. Murray and J. von Neumann, *On rings of operators II*, Trans. Amer. Math. Soc., 41, (1937), 208–248.

[MvN3] F. Murray and J. von Neumann, *On rings of operators IV*, Ann. Math., 44, (1943), 716–808.

[Ocn] A. Ocneanu, *Quantized groups, string algebras and Galois theory for algebras,* Operator Algebras and Appl., Vol.2 (Warwick 1987), London Math. Soc. Lecture Notes Ser. Vol. 136, pp. 119–172, Cambridge University Press, 1988.

[OK] A. Ocneanu (Lecture Notes written by Y. Kawahigashi), *Quantum Symmetry, Differential Geometry of Finite Graphs, and Classification of Subfactors,* Univ. of Tokyo Seminar Notes, 1990.

[PP1] M. Pimsner and S. Popa, *Entropy and index for subfactors*, Ann. Sci. Ec. Norm. Sup., 19, (1986), 57–106.

[PP2] M. Pimsner and S. Popa, *Iterating the basic construction,* Trans. Amer. Math. Soc., 310, (1988), 127–133.

[Pop1] S. Popa, *Maximal injective subalgebras in factors associated with free groups,* Adv. in Math., 50, (1983), 27–48.

[Pop2] S. Popa, *Orthogonal pairs of subalgebras in finite von Neumann algebras,* J. Operator Theory, 9, (1983), 253–268.

[Pop3] S. Popa, *Relative dimension, towers of projections and commuting squares of subfactors,* Pacific J. Math., 137, (1989), 181–207.

[Pop4] S. Popa, *Classification of subfactors: the reduction to commuting squares,* Invent. Math., 101 (1990), 19–43.

[Pop5] S. Popa, *Sur la classification des sous-facteurs d'indice fini du facteur hyperfini,* Comptes Rendus Acad. Sci. Paris, Sér. I, 311, (1990), 95–100.

[Pop6] S. Popa, *Classification of amenable subfactors of type II,* Acta Math., 172, (1994), 163–255.

[Pow] R. Powers, *Representations of uniformly hyperfinite algebras and their associated von Neumann rings*, Ann. Math., 86, (1967), 138–171.

[Sak1] S. Sakai, *C^*-algebras and W^*-algebras,* Springer-Verlag, 1983.

[Sak2] S. Sakai, *A characterisation of W^*-algebras*, Pacific J. Math., 6, (1956), 763–773.

[SY] J. Seberry and Yamada, *Hadamard matrices, sequences and block designs*, Contemporary Design Theory, ed. J.H. Dinitz and D.R. Stinson, pp. 431–560 (1992).

[Seg] I.E. Segal, *Irreducible representations of operator algebras,* Bull. Amer. Math. Soc., 53, (1947), 73–88.

[Sk] C. Skau, *Finite subalgebras of a von Neumann algebra*, J. Functional Anal., 25, (1977), 211–235.

[Sun1] V.S. Sunder, *An invitation to von Neumann algebras*, Springer-Verlag, 1987.

[Sun2] V.S. Sunder, *Pairs of II_1 factors*, Proc. Indian Acad. of Sci., (Math. Sci.), 100, (1990), 357–377.

[Sun3] V.S. Sunder, *II_1 factors, their bimodules and hypergroups*, Trans. Amer. Math. Soc., 330, (1992), 227–256.

[Sun4] V.S. Sunder, *A model for AF algebras and a representation of the Jones projections*, J. Operator Theory, 18, (1987), 289–301.

[SV] V.S. Sunder and A.K. Vijayarajan, *On the non-occurrence of the Coxeter graphs β_{2n+1}, E_7 and D_{2n+1} as principal graphs of an inclusion of II_1 factors*, Pac. J. Math., 161, (1993), 185–200.

[Tak1] M. Takesaki, *Theory of operator algebras I,* Springer-Verlag, 1979.

[Tak2] M. Takesaki, *Tomita's theory of modular Hilbert algebras and its applications*, Springer Lect. Notes in Math., 128, (1970).

[vN1] J. von Neumann, *On rings of operators III*, Ann. Math., 41, (1940), 94–161.

[vN2] J. von Neumann, *On infinite direct products*, Compos. Math., 6, (1938), 1–77.

[vN3] J. von Neumann, *On rings of operators. Reduction theory,* Ann. Math., 50, (1949), 401–485.

[Was] A. Wassermann, *Coactions and Yang–Baxter equations for ergodic actions and subfactors*, Operator Algebras and Appl., Vol.2 (Warwick 1987), London Math. Soc. Lecture Notes Ser. Vol. 136, pp. 202–236, Cambridge University Press, 1988.

[Wen] H. Wenzl, *Hecke algebras of type A_n and subfactors*, Invent. Math., 92, (1988), 349–383.

[Zim] R. Zimmer, *Ergodic theory and semisimple Lie groups*, Birkhäuser, 1984.

Bibliographical Remarks

§1.1: Almost all the material in this section was first established in the seminal paper [MvN1]. The only exceptions to this blanket statement is the existence of a tracial state on a finite factor (Proposition 1.1.2) – which was established in [MvN2] – and the fact (briefly alluded to) about disintegration of a general von Neumann algebra (with separable pre-dual) into factors – which was established in [vN3].

§1.2: The fundamental GNS construction first made its appearance in [GN], and appears later in polished form in [Seg], while the basic facts contained in this section concerning the so-called standard module of a finite factor were all known to the founding fathers – see [MvN2] (although they did not quite use the same terminology as here).

§1.3: The notion of a discrete crossed product – at least when the algebra that is being acted upon by the group is commutative – first appears in [MvN1]; already in this paper, they use this construction to give examples of II_1 factors, and they identify ergodicity as the crucial property of the group action to ensure factoriality of the crossed product.

§1.4: The definition given in Definition 1.4.2 is from [Kal]. The type-classification given in Theorem 1.4.5 is again from [MvN1]. The Powers factors made their appearance in [Pow], while the description of the model for the hyperfinite II_∞ factor coming from the crossed product $L^\infty(\mathbb{R}^2, \mathcal{B}, \mu) \times SL(2, \mathbb{Z})$ (in Example 1.4.8) is due to [Aub]. Infinite tensor products were first treated in [vN2], and while the uniqueness statement concerning approximately finite-dimensional II_1 factors was first proved in [MvN3], the ultimate classification of (all types of) approximately finite-dimensional factors was completed – except for one case, the so-called III_1 case – in [Con1]; the outstanding III_1 case was finally disposed of in [Haa].

§2.1: The classification of all possible modules over a factor goes back to [MvN3].

§2.2: The importance of bimodules was first recognised by Connes [Con2]. The coupling constant was introduced in [MvN1]. All the assertions in Proposition 2.2.6 appear in [Jon1] although many of these can also be found in the papers of Murray and von Neumann.

§2.3: This entire section is from [Jon1].

§3.1: The basic construction appeared in [Ch] and [Sk], although it was not really exploited in the manner discussed here until [Jon1]. Further, the index of a subfactor was first considered in [Jon1], and indeed, most of this section is also from that source.

§3.2: The notion of a Bratteli diagram was first systematically used in [Bra]. Most of the discussion in this section is 'folklore'; thus, for instance, Lemma 3.2.2 may be found (in possibly different pieces) in [GHJ]. On the other hand, facts concerning the Markov trace – such as Proposition 3.2.3 and Corollary 3.2.5 – occur in [Jon1].

§3.3: But for the reference to Kronecker's theorem concerning integral matrices of small norm – which may be found in [GHJ] – all of this section is also from [Jon1].

§4.1: The importance of bimodules was identified and underlined in [Con1]; their significance for subfactors was recognised in [Ocn]; the treatment given here may be found in [Sun3].

§4.2: The importance of the principal graph invariant of a subfactor – at least in the 'relative commutant formulation' – was already recognised in the first paper [Jon1] on subfactors, where it was also shown that the A_n diagrams all arose as principal graphs. The 'bimodule formulation' of the principal graph is due to Ocneanu ([Ocn]). The fact that the principal graph had to be one of the Coxeter diagrams was recognised in [Jon1]. It was in [Ocn] that it was stated that E_7 and D_{2n+1} could not arise as the principal graph invariant of a subfactor; (independent) proofs of this fact were furnished in [Iz1] and [SV]. It was later shown, in [Kaw], [B-N] and [Iz2] respectively, that all the graphs D_{2n}, E_6 and E_8 did in fact arise as principal graphs. The principal and dual graphs of the 'subgroup-subfactor' – referred to in Example 4.2.3(iii) were explicitly computed in [KY]. The real significance of the usefulness of the so-called 'diagonal subfactor' discussed in Example 4.2.4 has been brought out by Popa (see [Pop6]), who, incidentally, also computed the principal graph of this diagonal subfactor.

§4.3: Everything in this section is from [PP1], the only exception being

Proposition 4.3.6 which is from [PP2]. (As explained in the text, we have worked here with a marginal variation of what they term an 'orthonormal basis' – in that we replace the requirement of 'orthonormality' by 'linear independence'.)

§4.4: Everything in this section is contained in the work of Ocneanu and Popa (if not in such explicit detail).

§5.1: Practically everything in this section is from [PP1]. The notion of a commuting square, however, appeared much earlier in the work of Popa – see [Pop1] and [Pop2].

§5.2: Much of this section is motivated by considerations in [Jon3] and [Jon5]. Hadamard matrices originated in [Had]; also see [SY].

§5.3: Lemma 5.3.1, Corollary 5.3.2 and Lemma 5.3.3 are from Wenzl's thesis – see [Wen]. The terminology 'symmetric commuting square' (in the case of finite-dimensional C^*-algebras) goes back to [HS]; the terminology 'non-degenerate' to describe the same notion (but for more general inclusions of von Neumann algebras) is due to Popa – see [Pop6], for instance.

§5.4: The contents of this section were independently obtained in [Ocn] and [Sun4].

§5.5: Essentially all of this section is contained – although in a seemingly different form – in [Ocn]. The formulation contained here – at least as far as Proposition 5.5.2 is concerned – is explicitly worked out in [HS].

§5.6: The basic theorem stated in Theorem 5.6.3 was announced, without proof, in [Ocn]. Subsequently, this was proved in full detail in [Pop4]; the ultimate formulation of this theorem – in terms of (strong) amenability – is in [Pop6].

§5.7: Ocneanu's compactness result – both parts of it, as stated here – is from [OK]. The assertion of Corollary 5.7.4(ii) was first established in [Wen], although the proof given here, using bases, is different (and perhaps simpler).

§6.1: The diagrammatic formulation of the higher relative commutants, for vertex and spin models, is due to the first author (unpublished notes).

§6.2: The computations of the principal graphs of the examples considered in Examples 6.2.1 and 6.2.2 were independently performed in [BHJ] and [KSV]. These sort of Cayley graphs had, of course, been obtained in more general situations, in considerations of 'diagonal subfactors' in [Pop5] (also see [Pop6]) and of subfactors arising as fixed-point algebras of compact group actions – see [GHJ] and [Was].

§6.3: The contents of this section come from [KS].

§A.1: The equivalence of the abstract and concrete notions of a von Neumann algebra was established by Sakai in [Sak2].

§A.2: For the Tomita–Takesaki theorem, see [Tak2].

§A.4: The use of the co-finite morphism and the bimodule calculus for computing the principal graph of the subfactor $N \subseteq N \times G$ was demonstrated in [OK]. The computation of the principal and dual graphs for the subfactor $N \times H \subseteq N \times G$ was carried out in [KY].

§A.5: Most of this material appeared first in [Jon4] – with the obvious exception of the results of Artin, Alexander and Markov, which appeared first in [Art], [Alex] and [Mark], respectively; for these results, also see [Bir].

Index